Disclaimer

The publisher of this book is by no way associated with the National Institute of Standards and Technology (NIST). The NIST did not publish this book. It was published by 50 page publications under the public domain license.

50 Page Publications.

Book Title: Characterization of the US Construction Labor Supply

Book Author: Stanley W. Gilbert

Book Abstract: This study characterizes the construction labor pool, and carries out preliminary work toward an estimation of the supply and demand for construction labor. Specifically, it evaluates the composition of the construction labor force by race, age, educational attainment, union membership and employer type, and how that composition is changing over time. It also identifies which industries are most closely related to construction, and estimates labor flows over time by race, place of birth, and age. Finally, the report makes a preliminary evaluation of how skills have changed in the construction labor force over time and how the skill level of the construction labor force changes with changing wages.

Citation: NIST SP - 1135

Keyword: Construction; Labor Supply; Econometric Analysis; Labor Flows; Related Industries

NIST Special Publication 1135

Characterization of the U.S. Construction Labor Supply

Stanley W. Gilbert

http://dx.doi.org/10.6028/NIST.SP.1135

NIST
National Institute of Standards and Technology
U.S. Department of Commerce

NIST Special Publication 1135

Characterization of the U.S. Construction Labor Supply

Stanley W. Gilbert
Applied Economics Office
Engineering Laboratory

http://dx.doi.org/10.6028/NIST.SP.1135

December 2012

U.S. Department of Commerce
Rebecca Blank, Acting Secretary

National Institute of Standards and Technology
Patrick D. Gallagher, Under Secretary of Commerce for Standards and Technology and Director

Certain commercial entities, equipment, or materials may be identified in this document in order to describe an experimental procedure or concept adequately. Such identification is not intended to imply recommendation or endorsement by the National Institute of Standards and Technology, nor is it intended to imply that the entities, materials, or equipment are necessarily the best available for the purpose.

National Institute of Standards and Technology Special Publication 1135
Natl. Inst. Stand. Technol. Spec. Publ. 1135, 98 pages (December 2012)
http://dx.doi.org/10.6028/NIST.SP.1135
CODEN: NSPUE2

Abstract

This study characterizes the construction labor pool, and carries out preliminary work toward an estimation of the supply and demand for construction labor. Specifically, it evaluates the composition of the construction labor force by race, age, educational attainment, union membership and employer type, and how that composition is changing over time. It also identifies which industries are most closely related to construction, and estimates labor flows over time by race, place of birth, and age. Finally, the report makes a preliminary evaluation of how skills have changed in the construction labor force over time and how the skill level of the construction labor force changes with changing wages.

Keywords:

Construction; Labor Supply; Econometric Analysis; Labor Flows; Related Industries

Preface

This study was conducted by the Applied Economics Office in the Engineering Laboratory at the National Institute of Standards and Technology. The study characterizes several aspects of the construction labor supply. The intended audience is the National Institute of Standards and Technology, as well as other government agencies, private sector organizations concerned about the perceived decline in construction productivity, and standards development organizations that produce standards used by the construction industry.

Disclaimer

Certain trade names and company products are mentioned in the text in order to adequately specify the technical procedures and equipment used. In no case does such identification imply recommendation or endorsement by the National Institute of Standards and Technology, nor does it imply that the products are necessarily the best available for the purpose.

Cover Photography Credits

Cover photos, clockwise from the upper left are: Construction workers looking at blue print, provided courtesy of Microsoft; Construction cranes against sky, provided courtesy of Fotolia; City construction site, provided courtesy of Microsoft; and construction site, provided courtesy of Microsoft.

Acknowledgements

The author wishes to thank all those who contributed so many excellent ideas and suggestions to this report. They include Dr. Robert Chapman, Chief of the Applied Economic Office (AEO) at the National Institute of Standards and Technology who provided support and many useful comments. Dr. David Butry, an AEO economist, provided many useful comments. Dr. James Filliben, of the Statistical Engineering Division at the National Institute of Standard and Technology provided comments that substantially improved the clarity and presentation of the results. Dr. Paul Goodrum of the University of Kentucky, College of Engineering reviewed an earlier version of this paper, and made several helpful suggestions. Ms. Carmen Pardo of the AEO assisted greatly in preparing the manuscript for review and publication. As always, any remaining flaws and errors are the responsibility of the author.

Table of Contents

Abstract iii
Preface v
Acknowledgements vii
Table of Contents ix
List of Figures x
List of Tables xii
Executive Summary xv
1. Introduction 1
2. Relationship to Other Industries 5
    2.1. Informal Theory and Methodology 5
    2.2. Results 6
3. Characteristics of the Construction Labor Pool 9
    3.1. Theory and Methodology 9
    3.2. Results 13
        3.2.1. Seasonality 16
        3.2.2. Union Membership 17
        3.2.3. Employer Type 19
        3.2.4. Cohort Effects on Employer Type and Union Membership 21
        3.2.5. Sex 24
        3.2.6. Age 27
        3.2.7. Race and Ethnicity 31
        3.2.8. Education 36
4. Labor Flows 39
    4.1. Model and Methodology 39
    4.2. White Men 42
    4.3. Black Men 47
    4.4. U.S. Born Hispanic Men 52
    4.5. Foreign Born Hispanic Men 57
5. Discussion 63
    5.1. Skills 63

    5.2. Labor Supply and Wage 65

    5.3. Future Directions 68

6. References 71

Appendices 73

    Appendix 1: Data 73

    Appendix 2: Similarity Indices 76

List of Figures

Figure 1: Labor Productivity index for the US Construction industry and all non-farm industries. 2

Figure 2: Interactions of the variables in the analysis. 13

Figure 3: Proportion of men in construction in each categorical variable in 2010. 14

Figure 4: Main-Effects Plot of Unemployment v. Key variables analyzed in the paper. 14

Figure 5: Cumulative Normalized Unemployment over the study period. 15

Figure 6: Average change in construction employment by month. 16

Figure 7: Percent union membership in construction estimated from the CPS (data points) and calculated from regression (line). Blue regions represent recessions. 18

Figure 8: Estimated and Calculated probability that a person in construction has a specified type of employer. 20

Figure 9: Changes in Employer type and Union membership with age for the 1960 cohort. 22

Figure 10: Changes in employer type and union membership with cohort at age 30. 22

Figure 11: Monthly male and female participation in the construction industry. Left scale is male participation and right scale is female participation. 24

Figure 12: Percent women employed in construction by month. 25

Figure 13: Monthly probability of being in construction given that a man is in a specified age category. 28

Figure 14: Monthly probability of being in a given age group given that a man is working in construction. 30

Figure 15: Monthly Employment Probability by race and ethnicity. 32

Figure 16: For each race/ethnicity, the fraction of men working in construction divided by the proportion of that race/ethnicity in the population, by month. 33

Figure 17: Monthly Conditional Probability by race and ethnicity. 34

Figure 18: Proportion of men in construction who belong to each race/ethnicity, by region and time. 35

Figure 19: Monthly Employment Probability by level of education. 37

Figure 20: Conditional Probability by educational level. 38

Figure 21: Blue line is the number of White Men in construction in 1994, by age. Red line is the stable number of White men in construction assuming economic growth and unemployment continue at their average rates indefinitely. 42

Figure 22: Fit of estimates to data for White men in construction by age group 44

Figure 23: Net inflow to construction for White men, by age for selected years. 45

Figure 24: Blue line is the number of Black Men in construction in 1994, by age. Red line is the stable number of Black men in construction assuming economic growth and unemployment continue at their average rates indefinitely. 47

Figure 25: Fit of estimates to data for Black men in construction by age group, by month. 49

Figure 26: Net inflow to construction for Black men, by age for selected years. 50

Figure 27: Blue line is the number of US Born Hispanic Men in construction in 1994, by age. Red line is the stable number of US Born Hispanic men in construction assuming economic growth and unemployment continue at their average rates indefinitely. 52

Figure 28: Fit of estimates to data for US Born Hispanic men in construction by age group. 54

Figure 29: Net inflow to construction for U.S. born Hispanic men, by age for selected years. ... 55

Figure 30: Blue lines both represent the number of foreign-born Hispanic Men in construction in 1994, by age. Red line is the stable number in construction assuming economic growth and unemployment continue at their average rates indefinitely. Note that the solid blue line is plotted against the left axis, while the red line (and dashed blue line for scale) is plotted against the right axis. 57

Figure 31: Fit of estimates to data for foreign-born Hispanic men in construction by age group. 59

Figure 32: Net inflow to construction for foreign-born Hispanic men, by age for selected years. 60

Figure 33: Average years of education of men in construction by age, and normalized by the average years of education of men in the population at large. 64

Figure 34: Average normalized years of education for 30-year-old men in construction by year. 65

Figure 35: Interpolated Monthly GDP overlaid on reported Quarterly GDP. 75

List of Tables

Table 1: Select 2009 Statistics for construction. Values are in Billions of Dollars unless otherwise stated. 1

Table 2: Top 12 industries in terms of similarity index with construction. 7

Table 3: Top 10 industries in terms of inter-industry flows with construction. 8

Table 4: List of variables analyzed below and the number of levels for each variable. 13

Table 5: Coefficients for the Conditional Probability estimate that a man employed in construction is a member of a union. 17

Table 6: Coefficients for the Conditional Probability that a man in construction has a specified type of employer. 19

Table 7: Marginal effects on the probability of working in the specified sector. 19

Table 8: Coefficients for the estimate of the Conditional Probability for employer type and union membership versus cohort and age. 21

Table 9: Coefficients for the estimate of Employment Probability by sex. 24

Table 10: Probability of being female given that a person is employed in construction. 25

Table 11: Coefficients for the estimate of Employment Probability by age. 27

Table 12: Coefficients for the estimate of the Conditional Probability by age. 29

Table 13: Marginal Effects on Conditional Probability by age group. 29

Table 14: Coefficients for the estimate of Employment Probability by race/ethnicity. 31

Table 15: Coefficients for the estimate of the Conditional Probability by race and ethnicity 33

Table 16: Marginal effects on Conditional Probability by race and ethnicity. 33

Table 17: Coefficients for the estimate of Employment Probability by level of education 36

Table 18: Coefficients for the estimate of the Conditional Probability by educational level. 37

Table 19: Marginal effects for Conditional Probability by educational level. 38

Table 20: Coefficients for net flows of White men into construction. 43

Table 21: Net rate of inflow to construction by age and rate of economic growth. Colored cells indicate ages and circumstances where men are on net leaving construction. 45

Table 22: Impact of seasonality. Seasonal Coefficient is the absolute magnitude of the seasonal variation. Wald Test represents the probability that the seasonal variation is greater than zero. Percent change is calculated based on the highest estimated seasonal employment compared to the lowest estimated seasonal employment. 46

Table 23: Coefficients for net flows of Black men into construction... 48

Table 24: Net rate of inflow to construction by age and rate of economic growth. Colored cells indicate ages and circumstances where men are on net leaving construction. 50

Table 25: Impact of seasonality. Seasonal Coefficient is the absolute magnitude of the seasonal variation. Wald Test represents the probability that the seasonal variation is greater than zero. 51

Table 26: Coefficients for net flows of US Born Hispanic men into construction........................ 53

Table 27: Net rate of inflow to construction by age and rate of economic growth. Colored cells indicate ages and circumstances where men are on net leaving construction. 55

Table 28: Impact of seasonality. Seasonal Coefficient is the absolute magnitude of the seasonal variation. Wald Test represents the probability that the seasonal variation is greater than zero. 56

Table 29: Coefficients for net flows of US Born Hispanic men into construction........................ 58

Table 30: Net rate of inflow to construction by age and rate of economic growth. Colored cells indicate ages and circumstances where men are on net leaving construction. 61

Table 31: Impact of seasonality. Seasonal Coefficient is the absolute magnitude of the seasonal variation. Wald Test represents the probability that the seasonal variation is greater than zero. 61

Table 32: Simultaneous estimation of construction wage and choice to work in construction for adult men. 67

Executive Summary

Construction is an engine of growth for the U.S. economy. Investment in plant and facilities, in the form of construction activity, provides the basis for the production of goods and the delivery of services. Investment in infrastructure promotes the smooth flow of goods and services and the movement of individuals. Investment in housing accommodates new households and allows existing households to expand or improve their housing. It is clear that construction activities affect nearly every aspect of the U.S. economy and that the industry is vital to the continued growth of the U.S. economy.

This study characterizes the construction labor supply, and in particular characterizes how it is changing over time. This is a preliminary step toward estimating the supply and demand for construction labor, which is itself part of an effort to understand changes in construction labor productivity.

It was not possible to characterize all variables that correlate with either construction labor supply or construction productivity due to limitations of both the data and time. The variables that were analyzed were chosen based on three main considerations. First, variables were selected based on data availability. Second, some variables were chosen because they have been previously identified as potentially being associated with the changes in construction productivity. Third, some variables are included because they may represent categories of workers that may respond differently to price signals in the construction labor market.

Four basic questions are answered in this study. First, the supply pool from which construction labor is drawn is identified. Second, composition of the work force is characterized, and how it changes over time. Third, net labor flows by age are estimated for several different groups within the construction labor force. Fourth, some specific issues related to skilled labor within the construction labor force are evaluated.

Data for this study are primarily taken from the Current Population Survey (CPS) from 1994 – 2010.

First, the supply pool from which construction labor is drawn is identified. In particular, the industries that are seen as approximate substitutes by construction workers, and those industries with relatively large labor flows between it and construction are identified. The main finding is that construction draws from a pool of industries that are low to medium skilled and not necessarily closely related to construction like retail trade and food-service. Closely related industries are usually either very small (in numbers of employees) or related to a specialty occupation within construction (like administrative and support services).

To evaluate the composition of the work force and how it changes over time, several characteristics are evaluated, including racial composition of the work force, which provides some insight into the extent to which there is an influx of unskilled foreign workers into the labor pool. In an effort to estimate skills and changing skill levels, the educational composition of the work force and changes in labor union membership are also evaluated. Finally, the age distribution of the workforce is evaluated.

The biggest change in the labor force is that Hispanics (probably dominated by the foreign-born) are growing rapidly. Union membership is declining. That means that the market for skilled trades is changing. The nature of those changes, however, cannot be determined from this data set. Lower skilled employees are the most susceptible to the business cycle. That includes non-union members, younger and less educated workers, and Hispanics.

Detailed findings include:

- The construction market is seasonal, with peak employment in the summer.

- About 10 % of the construction labor force is female.

- Union membership is declining at an average rate of about 2.5 % per year. Union membership is slightly countercyclical, which suggests that union members are slightly less susceptible to the business cycle[1] than non-union members. However, it is possible that the difference in susceptibility to the business cycle is due to demographic characteristics (like age differences) rather than union membership.

- Over the long term, private employment and self-employment in construction are increasing as a proportion of total employment, while government employment in construction is decreasing. In fact, government employment in the construction sector is decreasing in absolute terms. The business cycle affects private employment more than the other sectors.

- As privately employed non-union members get older, many of them move into self-employment. A similar process may be affecting privately employed union members as well. Publically employed union members are a small but constant fraction of construction workers, suggesting that such workers tend to stay put as they get older.

- The decline in union membership appears to be primarily due to younger cohorts choosing not to join unions rather than to existing members dropping out.

- More recent cohorts appear to be less likely to be self-employed. An estimated 30 % of people born in 1960 were likely to be self-employed at age 50, while only an estimated

---

[1] The business cycle "refers to economy-wide fluctuations in production or economic activity over several months or years" (Business cycle. (2011, October 3). In *Wikipedia, The Free Encyclopedia*. Retrieved 13:25, October 4, 2011, from http://en.wikipedia.org/w/index.php?title=Business_cycle&oldid=453740220). The business cycle is to be distinguished from seasonality (which is also analyzed here). Seasonality is the regular fluctuation for different times of year, while the business cycle typically spans several years and includes recession and recovery.

20 % of those born in 1990 will be self-employed at age 50. However, the shift in self-employment probability is a reflection of the changes in the racial and educational composition of the construction work-force. What is not clear is whether the changes in self-employment probability are caused by the demographic changes, or whether they are caused by other factors that correlate with the demographic changes.

- Susceptibility to the business cycle decreases with age. The oldest groups are nearly immune to the business cycle.

- Hispanic men are entering construction at an increasing rate compared to white or black men, while the likelihood of black men in construction is actually decreasing. Nevertheless, White men still make up about 75 % of the construction labor force.

- Hispanic men are more susceptible to the business cycle than black men, who in turn are more susceptible to the business cycle than white men.

- Initially, those with a high school education were the most likely to be in construction, with men without a high school education second most likely. However, the likelihood of a man without a high school education being in construction has increased at a faster rate than the likelihood for a man with a high school education.

- However, since the number of people with a high-school education greatly outnumbers those without one, the number of people in construction with a high school education still outnumbers those without one.

- The greater the level of education, the less susceptible a person is to the business cycle.

Labor flows were estimated to evaluate issues regarding the aging of the work force, the number of young people entering the industry, and shifts in the work force toward foreign immigrants. The basic questions answered here is who is entering the work force, who is leaving it, and when.

The bulk of entrants to the work force are younger than 25. In most years, young people entering the industry outnumber older people leaving it. So concerns about gentrification of the industry do not seem to be reflected in the data. Specific findings include:

- The bulk of the inflow of white men to construction over the long term occurs before the age of 21. In general, older worker are less susceptible to the business cycle than younger workers. Above the age of 55, the impact of the business cycle has no statistical effect on employment.

- The bulk of the inflow of black men to construction over the long term occurs before the age of 24. As usual, older worker are less susceptible to the business cycle than younger workers. Above the age of 55, the impact of the business cycle has no statistical effect on employment.

- The bulk of the inflow of US-born Hispanic men to construction over the long term occurs before the age of 25. As usual, older workers are less susceptible to the business cycle than younger workers. Above the age of 55, the impact of the business cycle has no statistical effect on employment.

- The rate of inflow of foreign-born Hispanic men to construction is decreasing with age. However, inflow is still significant to the age of 55. The bulk of the inflow of foreign-born Hispanic men to construction over the long term occurs before the age of 34. Above the age of 55, the impact of the business cycle has no statistical effect on employment.

There is a perception that the construction industry has difficulty attracting and retaining skilled workers, and as a result faces a shortage of skilled workers. This problem is exacerbated by a 30-year decline in real construction wages relative to workers in other industries. This raises a number of economic questions that this report was intended, in part, to address.

In competitive markets, shortages are resolved by increases in price. That raises the question of why wages haven't adjusted for the decline. If construction costs (including delay costs) have increased, why haven't wages? To address this, two questions were evaluated: to what extent can a decline in skills be discerned in the data; and how does labor supply adjust to changes in wage?

- There is some support for the idea that there is a decline in skill level among the construction labor force. Average normalized years of education for men in construction at age 30 seems to decrease over time. Educational level also seems to be inversely related to the business cycle. However, since a formal model of education was not specified, these observations cannot be statistically tested.

- Preliminary efforts to model supply and demand could not be statistically estimated. The most likely reason for the failure is that an omitted variable correlates with both wage and employment. The best candidate for such an omitted variable would be skill. If wages correlate with skill (as seems reasonable) and if low-skill people are the last hired and first fired, then the efforts to estimate supply and demand without taking into account skill will fail.

- When estimating labor supply directly from microeconomic data, the model finds that correlation between the construction wage and the choice to work in construction is negative. That implies that increases in construction wages are associated with people selecting *out* of construction. That suggests that the skill premium is higher in other industries than in construction. So (all else equal) there are relatively few 'highly skilled' construction workers because such people can earn more in other industries—presumably because they are more productive there.

There are a number of additional directions that would contribute to understanding the construction labor market.

- Characterization at the regional / local level.

  This report characterized labor supply at the national scale for the most part. However, construction is primarily a local market and there will be aspects of the market that will be obscured by looking at it nationally. For example, racial makeup (and probably seasonality) clearly differs from region to region. So deepening the analysis to look at the data at a regional scale would likely improve our understanding of the market.

- Supply and demand need to be estimated.

  Estimating supply and demand functions for construction labor would help. That turns out to be surprisingly difficult due to the high correlation between wages and employment. During times of increasing employment, wages (presumably) increase, but the people hired are at the low end of the wage scale while the people at the high end of the wage scale are susceptible to poaching by other industries. Times of decreasing employment present the reverse situation. That makes it difficult to tease out the relationship between supply and wage holding all else constant. Completing the task of estimating supply and demand will help fill in some of the missing pieces of the picture of the construction labor market.

- Labor Unions

  The perceived shortage in skilled labor is probably linked to the declines in union membership. So to understand what is going on, more information is needed on the place of unions in the market, why market share is declining (both from the supply side with people choosing whether to join and from the demand side of builders choosing whether to hire union labor), and what (if anything) is replacing unions in the marketplace. So to better understand the nature of skilled-labor shortages (or lack thereof) requires an understanding of the changing place of the trade unions in the market.

- Wage trends for skilled craft workers v. general construction labor

  The CPS data used to generate this report is not detailed enough to distinguish skilled craft workers in construction from general laborers. It is possible that wages for general construction labor are declining while "shortages" for skilled craft workers are causing their wages to increase. This could explain how there are "shortages" of skilled craft workers alongside declining construction wages. One way of assessing this possibility would be to look at the long-term trend of wages for craft workers versus general construction labor. Such data does not exist in the CPS, so other data sources would have to be found.

- Labor flows by educational level

    Expanding the analysis of labor flows to address educational levels would provide additional insight into long-term changes in educational levels in people entering construction.

- Analysis by market segment

    Eventually, this work needs to be done for different segments of the construction market. Housing is such a large portion of the market that the results above are likely dominated by that segment of the market. But other segments will likely be different.

1. Introduction

Construction is an engine of growth for the U.S. economy. Investment in plant and facilities, in the form of construction activity, provides the basis for the production of products and the delivery of services. Investment in infrastructure promotes the smooth flow of goods and services and the movement of individuals. Investment in housing accommodates new households and allows existing households to expand or improve their housing. It is clear that construction activities affect nearly every aspect of the U.S. economy and that the industry is vital to the continued growth of the U.S. economy.

The construction industry's contribution to gross domestic product (GDP) and employment in 2009 is shown in Table 1.[2] In 2009 construction was still in decline following the 2007 to 2009 recession, and so the contribution of construction to GDP and employment was substantially below its long-term average.

Table 1: Select 2009 Statistics for construction. Values are in Billions of Dollars unless otherwise stated.

Statistic	Value (billions)
Value of Construction put in place	937.21
Residential	28 %
Commercial / Industrial	35 %
Manufacturing	8 %
Public Works	29 %
US GDP	14,119.04
Construction Value Added	537.46
Percent of GDP	3.8 %
Total US Employment (millions)	137.775
Construction Employment	9.702
Percent of Total Employment	7.0 %

In spite of its importance to the economy, construction seems to be undergoing a long-term decline in productivity[3]. To illustrate, Figure 1 shows the results of Teicholz[4] for construction productivity growth over the last 40 years. He found that as measured by constant contract dollars of new construction work per field work hour, labor productivity in the construction industry has trended downward over the past 40 years at an average compound rate of -0.6 %

---

[2] Thomas, D., 2010. "Methodology for Calculating Construction Industry Supply Chain Statistics." NIST Special Publication 1116. Gaithersburg, MD.
[3] Not all researchers believe that construction productivity is in decline. For a discussion of the debate, see Huang, Allison, Robert Chapman and David Butry. (2009) *Metrics and Tools for Measuring Construction Productivity: Technical and Empirical Considerations.* NIST Special Publication 1101. Gaithersburg, MD.
[4] From Teicholz, Paul. "Labor Productivity Declines in the Construction Industry: Causes and Remedies." *AECbytes Viewpoint.* Issue 4. April 14, 2004.

per year. Reasons for the decline are debated, but some things that have been suggested as possible reasons include[5] a shortage of skilled workers (and in particular skilled craft workers), an aging work force, the possibility that fewer young people are entering the industry, and the influx of unskilled labor from abroad.

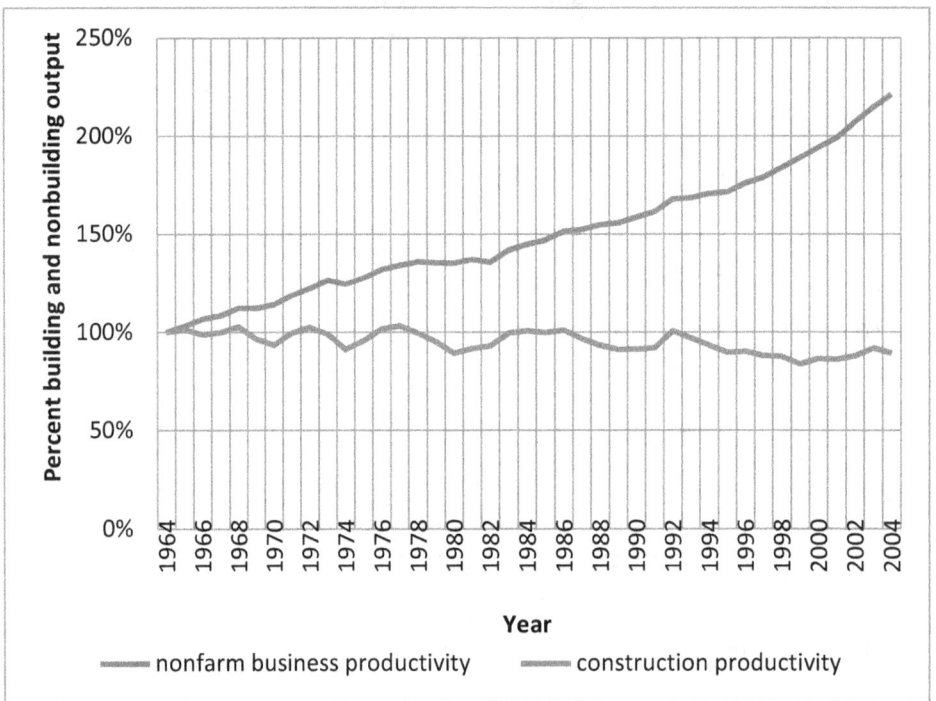

Figure 1: Labor Productivity index for the US Construction industry and all non-farm industries.[6]

This study is intended as a preliminary step toward understanding the changes in construction labor productivity. Since many of the factors believed to influence the changes in construction productivity are related to changes in the construction labor supply, this study characterizes construction labor supply, and in particular characterizes how it is changing over time. What it does not do is characterize how these factors influence labor productivity. That is the subject of future research.

It was not possible to characterize all variables that correlate with either construction labor supply or construction productivity due to limitations of both the data and time. The variables that were analyzed were chosen based on three main considerations. First, variables were selected based on data availability. Construction sector is almost certainly a significant factor in both labor supply and construction productivity. However, the data set used here contains no information on the construction sector a worker is in. Second, some variables were chosen

---

[5] Among other possible causes. For a more complete discussion of possible causes of the decline in productivity see Huang, Allison, Robert Chapman and David Butry. (2009) *"Metrics and Tools for Measuring Construction Productivity: Technical and Empirical Considerations."* NIST Special Publication 1101.
[6] Source: Teicholz, Paul. "Labor Productivity Declines in the Construction Industry: Causes and Remedies." *AECbytes Viewpoint.* Issue 4. April 14, 2004.

because they have been previously identified as potentially being associated with the changes in construction productivity. Education and labor union membership are included because they are potentially associated with the perceived shortage of skilled craft workers. Third, some variables are included because they may represent categories of workers that may respond differently to price signals in the construction labor market. Race and sex are included in this category.

This report analyzes the entire labor force for the construction industry (including construction workers, skilled tradesmen, office staff, management, etc.). In particular, this report provides details on the methodology used to obtain the results, and a detailed discussion of what can be concluded from the data.

Data for this study are taken from the Current Population Survey (CPS) from 1994 to 2010. Variables used are listed in Appendix 1. The CPS is a monthly survey of about 60 000 occupied households. The households represent a stratified random sample of households in the country. Each household is surveyed eight times in two blocks. A household is first surveyed for four consecutive months. The household is then out of the survey for the following eight months, followed by a second four-month period when it is surveyed. With this design, consecutive months have a 75 % overlap in the households in the survey, and surveys one year apart have a 50 % overlap in the households in the survey. A total of 18 variables from the data set are included in the analysis. The dataset contains a total of about 27 million records. Adult men employed in construction in the data set number about 800,000.

There are matching issues with the longitudinal aspect of the data. In particular, the months of July 1995, March 2000, and March 2001 people in the survey cannot be matched with the surrounding months. However, this seems unlikely to bias the results where longitudinal elements are used.

A number of factors suggest that the data before the middle of 1995 may not be as reliable as the rest of the data. Reasons for such a conclusion are discussed in Appendix 1.

Definitions of industries used in the CPS changed between December 2002 and January 2003. That makes industry comparisons possibly problematic for the time periods 1994 to 2002 and 2003 to 2010. Analysis of the data and definitions suggests that for construction data from the two periods is roughly comparable. However, for some other industries that is certainly not the case.

Supplemental data for quarterly GDP were obtained from the Bureau of Economic Analysis. Monthly data for Unemployment and Urban Consumer Price Index (CPI-U) were obtained from the Bureau of Labor Statistics. For purposes of this report, data was collected from 1993 to 2010. Since the main data are monthly and GDP are published quarterly, it is necessary to interpolate the GDP data to monthly to effectively make use of it. Interpolation methods are discussed in Appendix 1.

Section 2 provides a measure of how closely related construction is to other industries, and how large the labor flows are between other industries and construction. Section 3 provides a detailed discussion of the characteristics of the construction labor pool. Section 4 evaluates the nature and magnitude of labor flows in and out of construction by worker characteristic. Section 5 concludes by discussing implications of the results, and suggesting directions for future research.

2. Relationship to Other Industries

In characterizing construction labor supply, it is helpful to identify the supply pool from which construction labor is drawn. In particular, it would be helpful to identify industries that are seen as approximate substitutes by construction workers. As an additional objective, identifying industries where cross-price elasticities might be significant is an important part of estimating supply and demand for construction labor.

2.1. Informal Theory and Methodology

One natural way to identify similar industries is to look at job changes. If two industries are seen as perfect substitutes by workers, then a person in one industry changing jobs will be relatively likely to move into the other industry.[7]

Let J be the set of industries, and for any $j \in J$, $|j|$ is the number of workers in that industry. Suppose $K \subset J$ is a set of industries seen as perfect substitutes, and $i, j \in K$ are two industries from that set. Consider a person in industry i who is changing jobs. That person will be equally likely to pick any available job from the set of all jobs available in the industries in K. If we assume that the number of jobs available is proportional to the number of people employed in each industry, then the probability (P) of persons in industry i who change jobs switching to industry j is:

$$P\{j|i\} = q \frac{|j|}{|K|}$$

Where $q = P\{K | i\}$, which is the probability of persons in industry i who change jobs switching to any industry found in set K. Rewriting, we get:

$$\frac{P\{j|i\}}{|j|} = \frac{q}{|K|} = \frac{P\{i|i\}}{|i|}$$

Where $P\{i|i\}$ is the probability of persons in industry i who change jobs remaining in industry i. In general, we expect different industries to be imperfect substitutes. So, we expect:

$$\frac{P\{j|i\}}{|j|} \leq \frac{P\{i|i\}}{|i|}$$

Where equality holds only in the case of perfect substitutes.[8]

---

[7] They would also need to be seen as perfect substitutes by employers as well. If they were not, then employers would be more likely to reject applicants from the other industry on the belief that their skills were not as good a match.

[8] Strictly speaking this analysis applies only in a (relatively) static world where numbers of jobs in each industry are not changing relative to each other. If, for example, you have two moderately closely related industries (but

That suggests a natural index of substitutability for construction jobs.

Define the index of substitutability as:

$$r_{ij} = \frac{|i \cap \{ji\}|}{|i \cap \{ii\}|}$$

Note that in general $r_{ij} \neq r_{ji}$. In fact, they may be quite different since the assumption of perfect substitutability does not usually hold.

The index above has one major limitation. An industry may be a perfect substitute for construction from the perspective of the labor market, but if it is very small its impact on the construction labor market will be small as well. So it would be useful to develop a second index that estimates impact on the labor market. Here, a very simple index can give a useful idea.

$$r_i = \frac{|i \cap \{ij\}| + |i \cap \{ji\}|}{2|i \cap \{ii\}|}$$

This represents the gross flows between construction and the compared industry normalized by the construction-to-construction job flows.

All terms are readily estimated from the CPS data. Since industry definitions change between 2002 and 2003, indexes are estimated solely for the period 2003 to 2010. Industry populations are the average population over the time period in question.

2.2. Results

Full results are listed in Appendix 2. The calculated similarity indices for the top 12 industries from the 2003 – 2010 CPS are listed in Table 2.

---

still imperfect substitutes) one of which is rapidly growing while the other is rapidly declining, you could have $\frac{\{ij\}}{|i|} > \frac{\{ji\}}{|j|}$ even though they are imperfect substitutes. Where the above analysis fails is in the assumption that the number of jobs available is proportional to the number of people employed. In a growing industry the ratio of jobs available to people employed will be relatively high, while in declining industries jobs available will be relatively scarce.

Table 2: Top 12 industries in terms of similarity index with construction.

Industry	r_{ij}
Wood products	17.04 %
Forestry, logging, fishing, hunting, and trapping	16.16 %
Petroleum and coal products manufacturing	15.45 %
Mining	15.12 %
Waste management and remediation services	14.85 %
Furniture and fixtures manufacturing	14.70 %
Nonmetallic mineral product manufacturing	14.24 %
Administrative and support services	13.80 %
Repair and maintenance	13.61 %
Agriculture	12.60 %
Utilities	10.57 %
Primary metals and fabricated metal products	10.11 %

Construction, like any industry, employs people from a number of different occupations. Top five occupations (as defined in the CPS) employed in construction are shown in a nearby table. It is worth noting that two of the occupations (Office and Administrative Support and Installation, Maintenance and Repair) match two of the industries on the similar industries list. That suggests that if the analysis were restricted to those occupations, the similarity indices for those industries would be much higher.

Calculated labor flow indices for the top 10 industries are listed in Table 3. Complete results are listed in Appendix 2. Unsurprisingly, there is very little overlap with the industries in Table 2. Many of the most similar industries are very small, and so flows between them will be small.

Top 5 Occupations Employed in Construction	
Construction and Extraction	66.48 %
Management	13.72 %
Office and Administrative Support	5.60 %
Installation, Maintenance and Repair	4.89 %
Transportation	2.67 %

Meanwhile, retail trade, which does not appear in the top 12 similar industries, has the highest inter-industry flows because it is by far the largest industry (in terms of employment) in the survey.

The results (mainly from Table 3) suggest that construction draws from a pool of industries that are low to medium skilled and not necessarily closely related to construction. Some occupations within construction (like administrative and support services) likely draw from a more specialized pool. However the relatively low-skill industries like retail trade and food-service make up a large portion of the industries that construction shares labor with.

Table 3: Top 10 industries in terms of inter-industry flows with construction.

Industry	f
Retail trade	8.93 %
Administrative and support services	6.61 %
Transportation and warehousing	4.66 %
Food services and drinking places	4.52 %
Professional and technical services	3.24 %
Wholesale trade	3.09 %
Repair and maintenance	2.81 %
Educational services	2.52 %
Agriculture	2.25 %

3. Characteristics of the Construction Labor Pool

It is believed that the changing composition of the work force may be a factor in the decline in construction productivity. In this section the composition of the work force is characterized, and how it changes over time. The educational composition of the work force and changes in labor union membership would help understand the skill level of construction labor and how it is changing. The age distribution of the workforce is also estimated.

3.1. Theory and Methodology

In this section, two questions are evaluated. First, what is the probability that a member of a specified group will be working in construction? For example, about 12 % of Hispanic men worked in construction in 2010. Second, given that a person is in construction, what is the probability that he is a member of that predefined group? For example, of all the men working in construction in 2010, about 20 % of them are Hispanic.

Let I be the set of all people (or in most cases all men) in the labor market. Then let $J = \{E_j\}$ be some partition of interest of the labor market—that is, for some $i, j \in J$, where $i \neq j$, $E_i \cap E_j = \emptyset$, and $I = \cup_j E_j$. The operator $\|E_j\|$ represents the total number of people in member j of the partition. In the work that follows the labor market is partitioned on the basis of sex, race and ethnicity, educational level, age, employer type, and union membership.

Consider the market for construction. The market can be specified implicitly with the following set of supply and demand equations:

$$0 = \tilde{D}_c(C, P; \alpha) = C - D_c(P; \alpha)$$
$$0 = \hat{S}_c(C, P, L, w; \beta) = C - S_c(P, L, w; \beta)$$
$$0 = \tilde{D}_c^j(L, P, L, w; \gamma) = L^j - D_c^j(C, P, L, w; \gamma)$$
$$0 = \hat{S}_c^j(L, w; \delta) = L^j - S_c^j(L, w; \delta)$$

Where

D_C	Aggregate demand for construction
S_C	Aggregate supply of construction
D_c^j	Aggregate demand for construction labor from partition j
S_c^j	Aggregate supply of construction labor from partition j
C	Total construction put in place
P	Price of construction
L	Vector of construction labor employed, where L^j is the total construction labor employed from partition j.
w	Vector of construction wages, where w^j is the construction wage for partition j.

X, Y, Z Exogenous factors, including among other things time, U.S. GDP (in X), prices of other factors (in Y), and unemployment rate (in Z).

Note that S_C and the q_i^c are in turn the joint solution to a production function for construction.

These implicit equations can be solved for the reduced-form equations of C, P, L, and w:

$$C^* = C(X, Y, Z)$$
$$P^* = P(X, Y, Z)$$
$$L^* = L(X, Y, Z)$$
$$w^* = w(X, Y, Z)$$

Of these functions, only $L(X, Y, Z)$ is estimated. As mentioned above, in this section $L(X, Y, Z)$ is estimated in two forms. First, $\frac{L_i}{\sum_j L_j}$ is estimated, then $\frac{L_i}{\sum_k L_k}$ is estimated for each member of the respective partitions. For simplicity of exposition, the first value will be referred to below as the Employment Probability, while the second will be referred to as the Conditional Probability.

Exogenous factors included in the estimations[9] are time, growth in U.S. GDP, and national unemployment.

If $\pi(t)$ represents the Employment Probability at time t, then the change in probability over time is:

$$\Delta \pi = \alpha \cdot \pi \Delta t + \beta \frac{\Delta Y}{Y} + \gamma \pi U \Delta t + o.t.$$

Where $\frac{\Delta Y}{Y}$ represents rate of economic growth, U represents current unemployment rate, α, β, and γ are parameters, and "o.t." represents other terms that will sometimes appear (e.g., seasonality). Letting $\Delta t \to 0$, we get:

$$\frac{1}{\pi} \frac{d\pi}{dt} = \alpha + \beta \frac{1}{Y} \frac{dY}{dt} + \gamma U + o.t.$$

Integrating gives:

$$\pi = \pi_0 \exp\left\{ \alpha t + \beta \ln Y + \gamma \int_{t_0}^{t} U \, d\tau + o.t. \right\}$$

This can be restated as rate of economic growth (instead of total economic activity):

---

[9] Except for the estimation of cohort effects, section 3.2.4.

$$y = y_0 \exp\left(\alpha t + \beta \ln \frac{y}{y_0} + \gamma \sum \cdots + \varepsilon \right)$$

or

$$y = y_0 \exp\left(\alpha t + \beta \ln(y+1) + \gamma \sum \cdots + \varepsilon \right)$$

or

$$y = \exp\left(\ln y_0 + \alpha t + \beta \ln(y+1) + \gamma \sum \cdots + \varepsilon \right)$$

Where g is annualized growth in U.S. GDP. Taking into account the fact that monthly growth and unemployment are stated in annualized terms, this then becomes:

$$y = \exp\left(\ln y_0 + \alpha t + \frac{1}{12}\beta \sum_{i=i_0}^{i} \ln(g_i+1) + \frac{1}{12}\gamma \sum_{i=i_0}^{i} u_i + \varepsilon \right)$$

$$y = \exp\left(\ln y_0 + \alpha t + \hat{\beta} \sum_{i=i_0}^{i} \ln(g_i+1) + \hat{\gamma} \sum_{i=i_0}^{i} u_i + \varepsilon \right)$$

The equation is normalized so that the parameter α represents average growth in the population over time. Specifically, if $\bar{g} = \frac{\sum \ln(g_i+1)}{n}$ and $\bar{u} = \frac{\sum u_i}{n}$, then estimation becomes:

$$y = \exp\left(\ln y_0 + \alpha t + \hat{\beta}\left(\sum_{i=i_0}^{i} \ln(g_i+1) - \bar{g}t\right) + \hat{\gamma}\left(\sum_{i=i_0}^{i} u_i - \bar{u}t\right) + \varepsilon \right)$$

Being in construction is a binary choice. As such the data are modeled using a binomial distribution. For any $i \in I$, $x_i = 1$ if the individual is in construction, and $x_i = 0$ if not. Then the likelihood function takes the form:

$$\prod_{i \in I} (p_i)^{x_i}(1-p_i)^{1-x_i}$$

Where

$$p_i = \exp\left(\ln p_0 + \alpha t_i + \hat{\beta}\left(\sum_{i=i_0}^{t_i} \ln(g_i+1) - \bar{g}t_i\right) + \hat{\gamma}\left(\sum_{i=i_0}^{t_i} u_i - \bar{u}t_i\right) + \varepsilon \right)$$

For a particular partition of the population $J = \{E_j\}$, the likelihood function becomes:

$$\ell\ell\ell = \prod \pi_j^{z_j}(1-\pi_j)^{1-z_j}$$

Where

$$\pi_j = \exp\left(\ln \pi_j^0 + \beta_j x_j + \hat{\epsilon}_j\right)\left(\ln(x_j+1) - \gamma g_j + \sum \delta \cdot z_j - \beta_j x_j + \epsilon_j\right)$$

which is estimated using Maximum Likelihood Estimation.

The analysis is restricted to the adult (age 16 or older) non-institutionalized population. For all but the results on sex below (Section 3.2.5), the analysis is further restricted to the male population.

Estimates of the Conditional Probability are calculated using standard multinomial logit. Specifically, if p_i^j is the probability that person i in construction is a member of group j, then:

$$p_i^j = \frac{\pi_i^j}{\sum_{j \in J} \pi_i^j}$$

Where the π_i^j are the same expressions as defined above (and no longer are interpreted as probabilities).

The expression is indeterminate without some normalization. The problem is solved by setting $\pi_i^j = 1$ for some $j \in J$.

All statistical analysis was performed using Stata 11 (StataCorp., 2009).

3.2. Results

Table 4 contains the list of the ten variables evaluated in this section. Figure 2 shows how the variables interact in the analysis below. In addition, there is one analysis in which the interaction of union membership with employee type, age, and cohort is examined. In that case, the business cycle variables (on the right hand side of the figure) are not included.

Table 4: List of variables analyzed below and the number of levels for each variable.

Variable	Number of levels
Union Membership	2
Employer type	4
Sex	2
Age	8
Race	3
Education	3
Year	(by month)
Seasonality	12 (months)
Growth in GDP	Continuous
Unemployment	Continuous

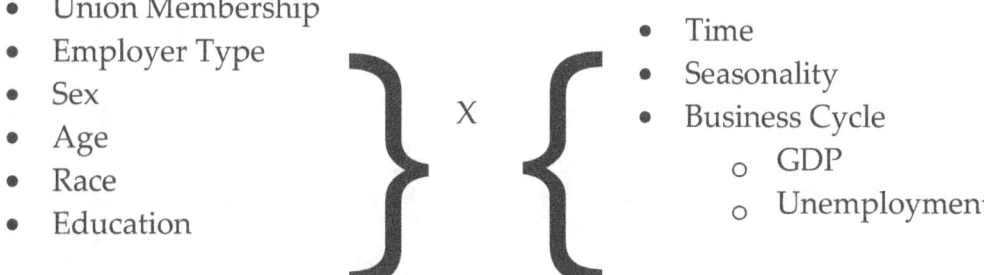

Figure 2: Interactions of the variables in the analysis.

Figure 3 shows the proportion of men employed in construction in each category of the categorical variables (except for sex) in 2010.

Employees in the construction industry tend to be white male, high-school educated, non-union, and employed in the private sector. Construction workers are almost all between the ages of 25 and 55.

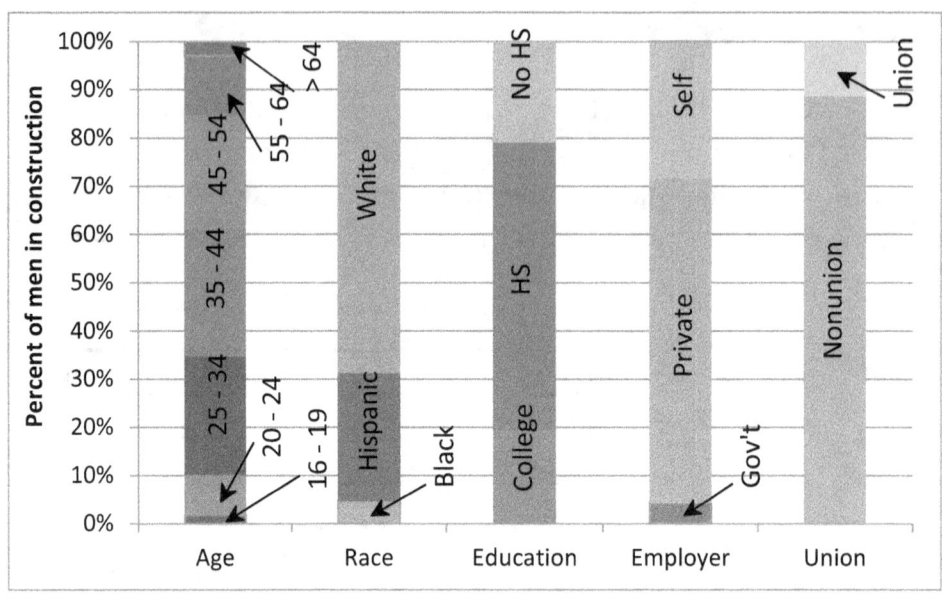

Figure 3: Proportion of men in construction in each categorical variable in 2010.

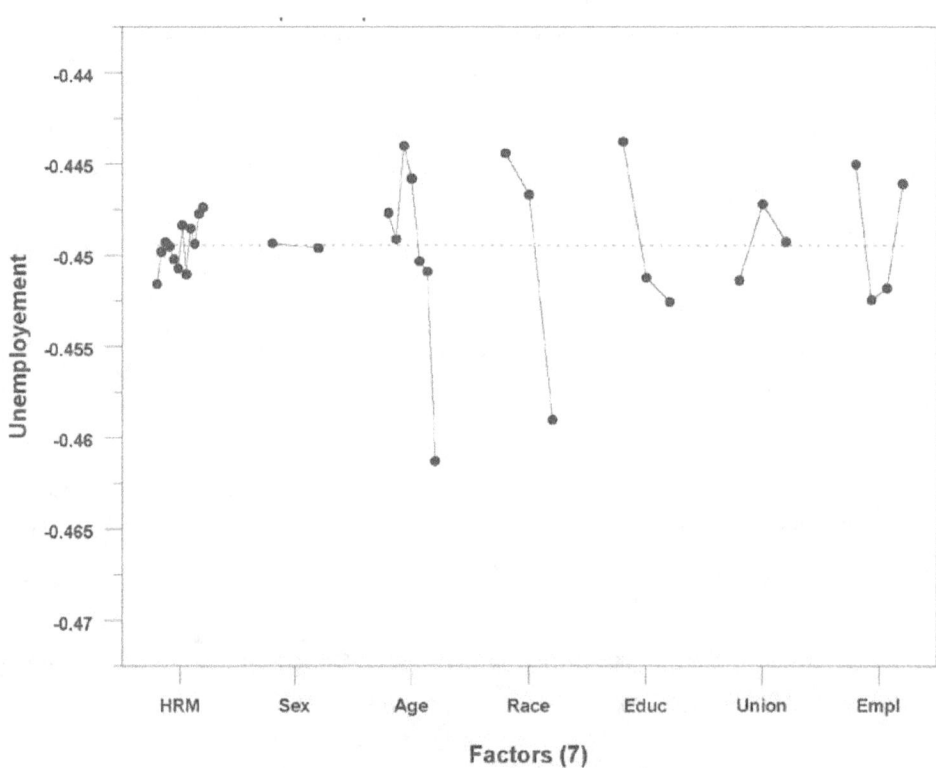

Figure 4: Main-Effects Plot of Unemployment v. Key variables analyzed in the paper.[10]

---

[10] HRM represents calendar month. For Sex, the first point represents males while the second represents females in the sample. The Age categories represent the same categories as in Figure 3. For Race, the categories in order are White, Black, and Hispanic. The Education categories in order are "No High School", High-School Education, and College. Union in order is Non-Union, Union, and unknown. Empl(oyer) categories in order are Unpaid, Self-Employed, Private, and Government.

Figure 4 is a Main-effects plot of unemployment versus key variables for people working in construction. Note that the unemployment term used in Figure 4 is the $\sum_{i=i_0}^{i_n} u_i - \bar{u}$ term from the regression above. As a comparison, Figure 5 shows the $\sum_{i=i_0}^{i_n} u_i - \bar{u}$ term plotted over the period of the study. If a group has low values in the plot, then the people who form that group tend to be in construction when the cumulative unemployment term is low. So, for groups that have low values in the plot, larger numbers are employed in construction later in the study period, and (secondarily) employment tends to correlate more strongly with the business cycle. Factors with large variations are those that distinguish between groups with large relative shifts in employment over time and those that have high vulnerability to the business cycle from those that have low vulnerability.

Based on this analysis, the sexes do not have large relative differences in employment over time, nor do they have large differences in vulnerability to the business cycle. Age and Race show large differences, with workers over 65 and Hispanics significantly more affected by the unemployment term. Results below suggest that the effect for age is probably primarily driven by changes over time, while the effects for Hispanics are due both to changes over time and vulnerability to the business cycle. The effects for education and unionization are likely due to a relative decline in the number of people without a high-school education and union members in construction over time. The impacts for Employer are probably primarily driven by vulnerability to the business cycle, with Self-Employed and Privately Employed people being more vulnerable to the business cycle than Government Employed people.

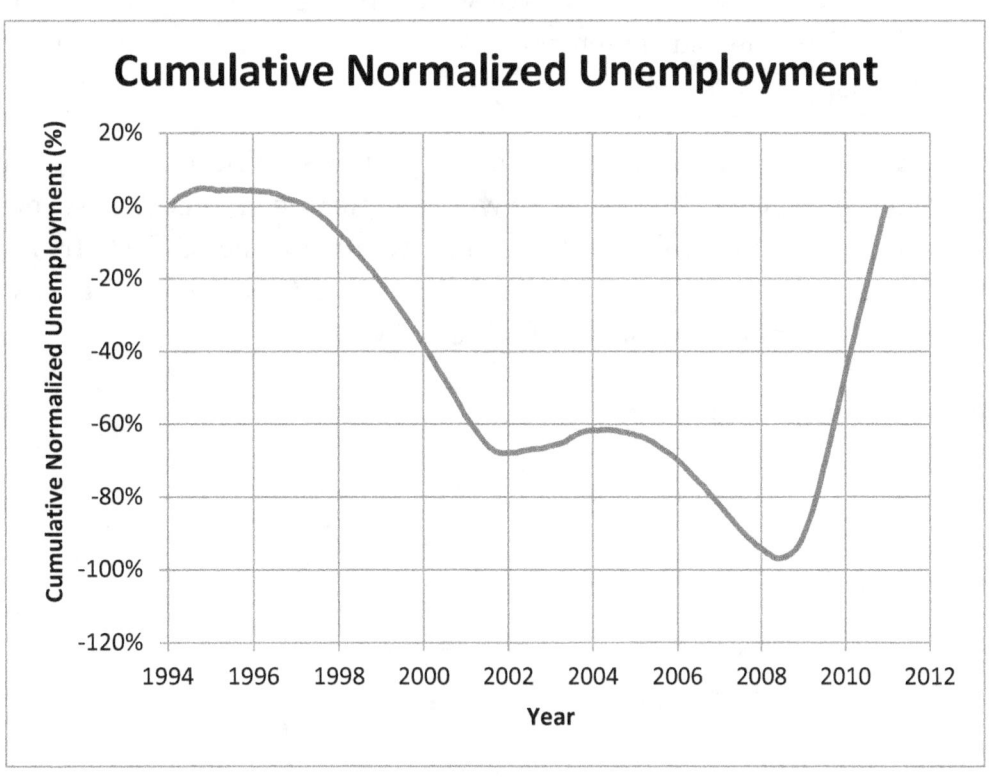

Figure 5: Cumulative Normalized Unemployment over the study period.

3.2.1. Seasonality

First, it is important to examine the effect of seasonality in construction employment. Monthly net change in construction employment is calculated and the results are averaged by month. These results are shown in Figure 6.

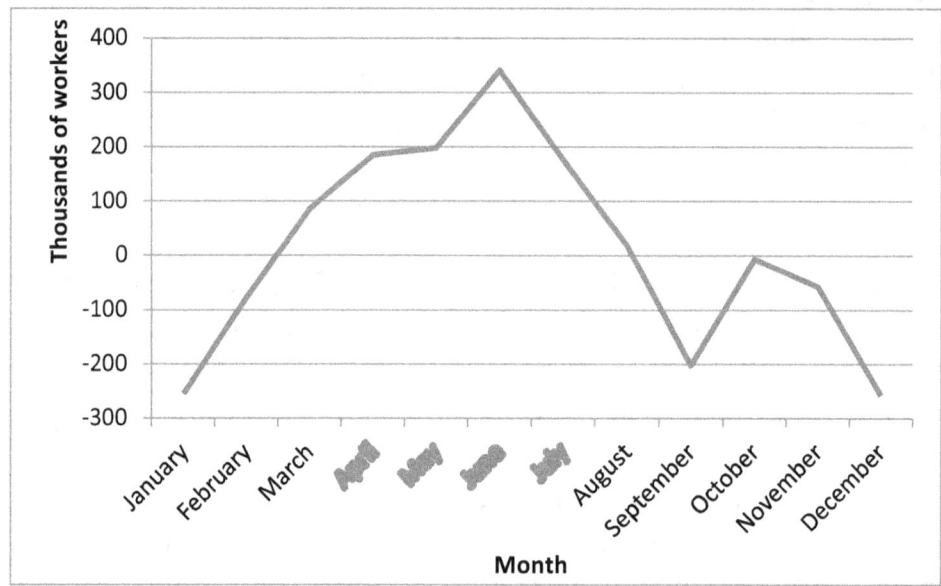

Figure 6: Average change in construction employment by month.

There is a clear seasonal effect on construction employment, with employment at its highest in summer. The June and September spikes are likely associated with the summer employment of students.

It seems likely that seasonal coefficients will vary significantly by region. The winter lull is probably strongest in the northeast where winter weather is most severe, while the seasonal effect is most likely weakest where winter weather is the mildest. It is possible that in areas where summers are the most severe (like the desert southwest) the seasonality could be reversed with employment at its highest in winter.

3.2.2. Union Membership

Results for the Conditional Probability regression for union membership are in Table 5. What is of interest are which variables are significant, and what is the magnitude and direction of the effect of each coefficient. The 'Pr' column gives the significance of a particular coefficient. Magnitude and direction of effects are discussed next.

Union membership is declining at an average rate of about 2.5 % per year. It goes from about 22.5 % in 1994 to about 15.5 % in 2010. Union membership is slightly countercyclical as demonstrated by the negative coefficient on growth. This suggests that union members are slightly less susceptible to the business cycle than non-union members. However, it is possible that the difference here is due to covarying demographic characteristics (like age and race differences) rather than union membership. Highest to lowest growth rate makes a difference to union membership of about 2 %. Highest to lowest unemployment rate also makes a difference of about 2 % in union membership.

Estimated versus calculated union membership is shown in Figure 7. The data points represent the weighted[11] percent of union members in the sample at each time, while the line is calculated from the results of the regression. In this figure and all subsequent figures, recessions are marked by a blue region. The vertical bars inserted in Figure 7, and in other figures throughout this report, designate periods of recession.

Table 5: Coefficients for the Conditional Probability estimate that a man employed in construction is a member of a union.

Var	Coef.	Std. Err.	z	Pr
years	-0.02508	0.002238	-11.2	< 0.001
growth	-0.08736	0.021251	-4.11	< 0.001
unemployment	-0.12796	0.038993	-3.28	0.001
Jan	-1.49016	0.023541	-63.3	< 0.001
Feb	-1.44587	0.02281	-63.39	< 0.001
Mar	-1.44993	0.023793	-60.94	< 0.001
Apr	-1.49228	0.023703	-62.96	< 0.001
May	-1.45917	0.023493	-62.11	< 0.001
Jun	-1.46885	0.022804	-64.41	< 0.001
Jul	-1.46903	0.022985	-63.91	< 0.001
Aug	-1.4139	0.022931	-61.66	< 0.001
Sep	-1.42104	0.022728	-62.52	< 0.001
Oct	-1.41976	0.022903	-61.99	< 0.001
Nov	-1.43575	0.023034	-62.33	< 0.001
Dec	-1.42685	0.022661	-62.96	< 0.001

---

[11] Weights in this section and throughout the report are supplied by the Census Bureau as part of the CPS data set and represent their best estimate as to how many people in the general population each person in the sample represents.

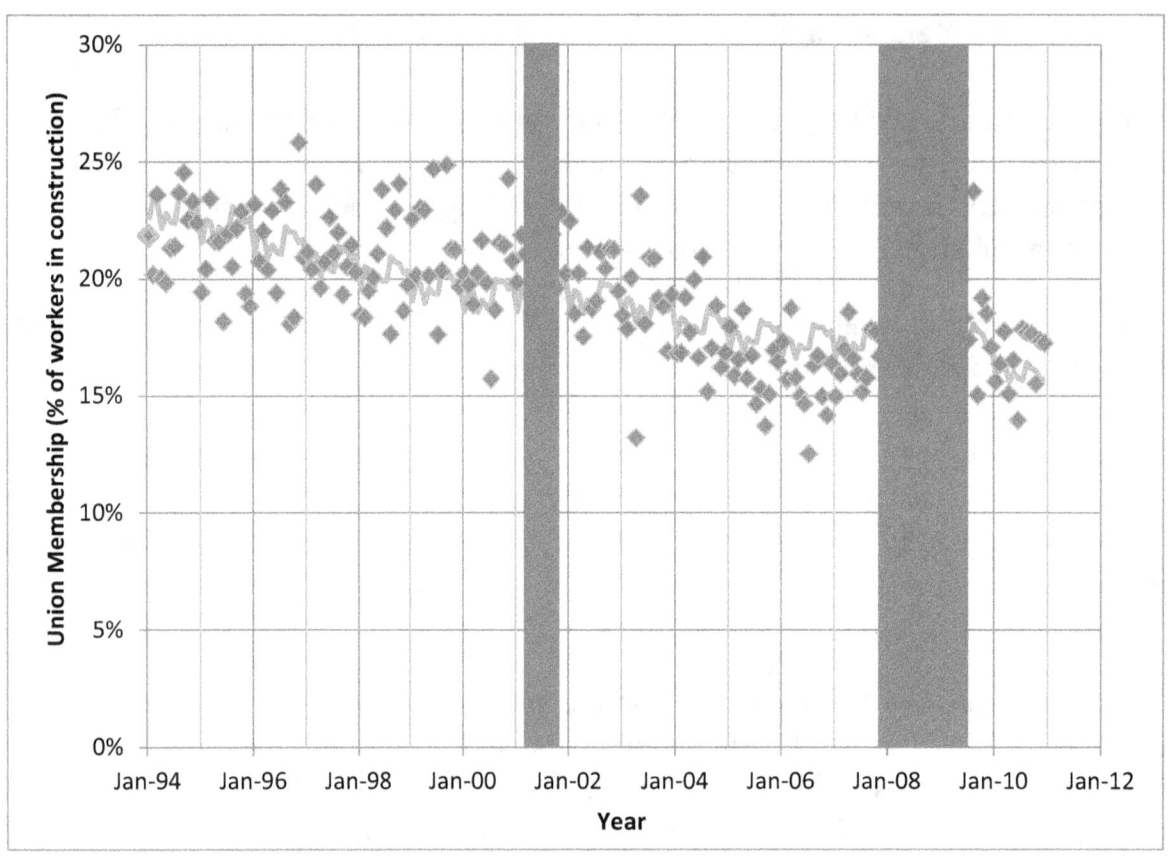

Figure 7: Percent union membership in construction estimated from the CPS (data points) and calculated from regression (line). Blue regions represent recessions.

3.2.3. Employer Type

Results for Conditional Probability by employer type are in Table 6[12]. A likelihood ratio test demonstrates that the unemployment variables are jointly not significantly different from zero ($\chi^2 = 2.77$, df = 3, p = 0.4286).

Table 6: Coefficients for the Conditional Probability that a man in construction has a specified type of employer.

	Coef.	Std. Err.	z	Pr
Government				
constant	-2.12879	0.021522	-98.91	< 0.001
years	-0.02672	0.003264	-8.19	< 0.001
growth	-0.8741	0.098955	-8.83	< 0.001
unemployment	-0.03937	0.187193	-0.21	0.833
Private (base outcome)				
Self-Employed				
constant	-0.85149	0.011999	-70.96	< 0.001
years	0.004958	0.001645	3.01	0.003
growth	-0.62204	0.050909	-12.22	< 0.001
unemployment	0.131137	0.093524	1.4	0.161
Unpaid				
constant	-5.59812	0.122181	-45.82	< 0.001
years	-0.07399	0.021849	-3.39	0.001
growth	-0.04714	0.589556	-0.08	0.936
unemployment	-0.90791	1.196151	-0.76	0.448

Results are shown in Figure 8.

Interpretation of the coefficients in Table 6 is not straightforward. To help in interpreting the results marginal effects are shown in Table 7. Marginal effects represent the effect on the respective types of employment by a unit change in one of the regression variables. For example, a one-year increase in time decreases government share of construction employment by 0.14 %.

Table 7: Marginal effects on the probability of working in the specified sector.

	Government	Private	Self	Unpaid
Years	-0.0014	0.0002	0.0014	-0.0001
Growth	-0.0359	0.1435	-0.1079	0.0002
Unemployment	-0.0038	-0.0211	0.0263	-0.0014

---

[12] As mentioned in the theory section, some normalization is required for the Conditional Probability estimation. In this case that is accomplished by setting all coefficients for people in Private employment to zero.

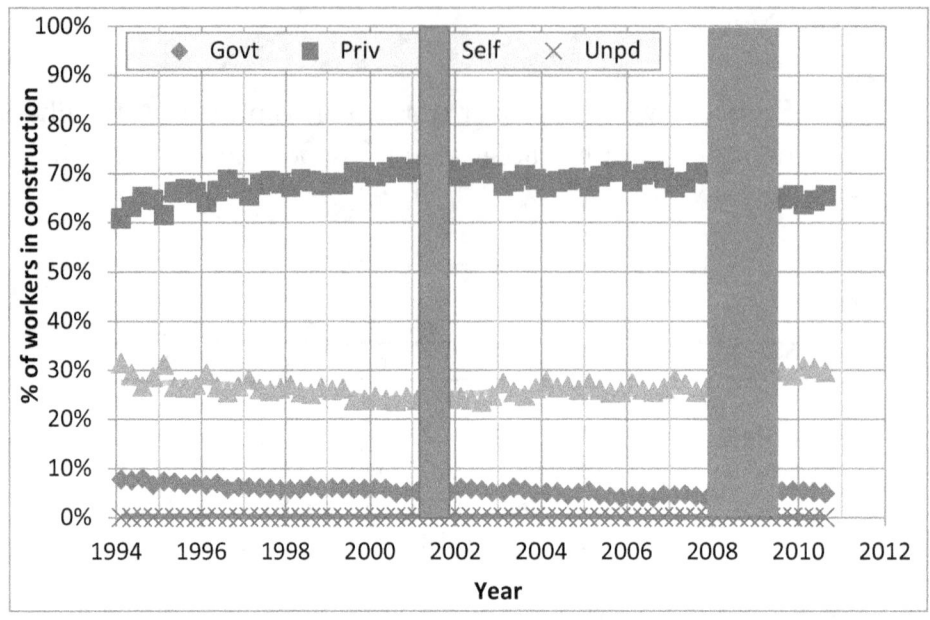

Figure 8: Estimated and Calculated probability that a person in construction has a specified type of employer.

Over the long term, private employment and self-employment are increasing as a proportion of total employment, while government employment is decreasing. In fact, government employment in the construction sector is decreasing in absolute terms (not shown here). According to the estimates, over the period of the study, government employment as a proportion of the construction industry has declined about 2 % while self-employment has increases by about the same amount.

The business cycle, as represented by the positive coefficient on growth and the negative coefficient on unemployment, affects private employment more than the other sectors. Highest to lowest growth rate makes a difference of about 7 % to the proportion of private employment with self-employment making up the majority of the difference. Highest to lowest unemployment rate makes a difference of about 1 % to the proportion of private employment with self-employment making up almost all the difference.

3.2.4. Cohort Effects on Employer Type and Union Membership

The Conditional Probability for combined employer and union membership are also evaluated. Here, though, the focus is on the effects of age and cohort on employer and union membership. Age is normalized as:

$$n\_age = \frac{age - 45}{40}$$

Cohort is defined as Year − Age, and is normalized as:

$$n\_cohort = \frac{cohort - 1960}{50}$$

In this case, the usual variables for time[13] and business cycle are not included. Results are listed in Table 8.

Table 8: Coefficients for the estimate of the Conditional Probability for employer type and union membership versus cohort and age.

	Coef.	Std. error	t value	Pr
Government × Union				
constant	-1.53508	0.043891	-34.9746	< 0.001
n_cohort	0.07445	0.537959	0.1384	0.890
n_age	-0.35404	0.445618	-0.7945	0.427
Government × Nonunion				
constant	-0.68792	0.021206	-32.4396	< 0.001
n_cohort	-0.1212	0.196962	-0.6153	0.538
n_age	1.597478	0.157668	10.1319	< 0.001
Private × Union				
constant	0.483497	0.020249	23.8775	< 0.001
n_cohort	-0.93809	0.142498	-6.5831	< 0.001
n_age	-0.70499	0.108587	-6.4924	< 0.001
Private × Nonunion				
constant	1.795408	0.019098	94.0079	< 0.001
n_cohort	0.333857	0.109615	3.0457	0.002
n_age	-0.80941	0.086114	-9.3992	< 0.001
Self-Employed				
constant	1.164634	0.020868	55.8105	< 0.001
n_cohort	-0.83631	0.124758	-6.7035	< 0.001
n_age	0.550928	0.095963	5.7411	< 0.001

The fitted estimates are shown in Figure 9 for the 1960 birth cohort. Results suggest that as privately employed non-union members get older, they move into self-employment. A similar process may be affecting privately employed union members as well. Publically employed

---

[13] Time is not included explicitly. However, since time = age + cohort, it is included implicitly.

union members are a small but constant fraction of construction workers, suggesting that such workers tend to stay put as they get older.

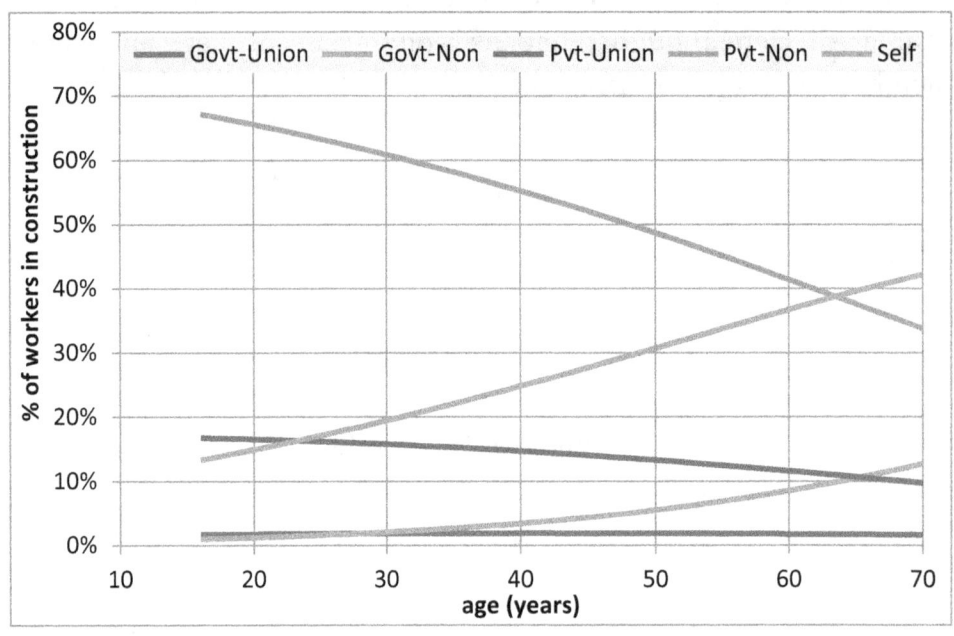

Figure 9: Changes in Employer type and Union membership with age for the 1960 cohort.

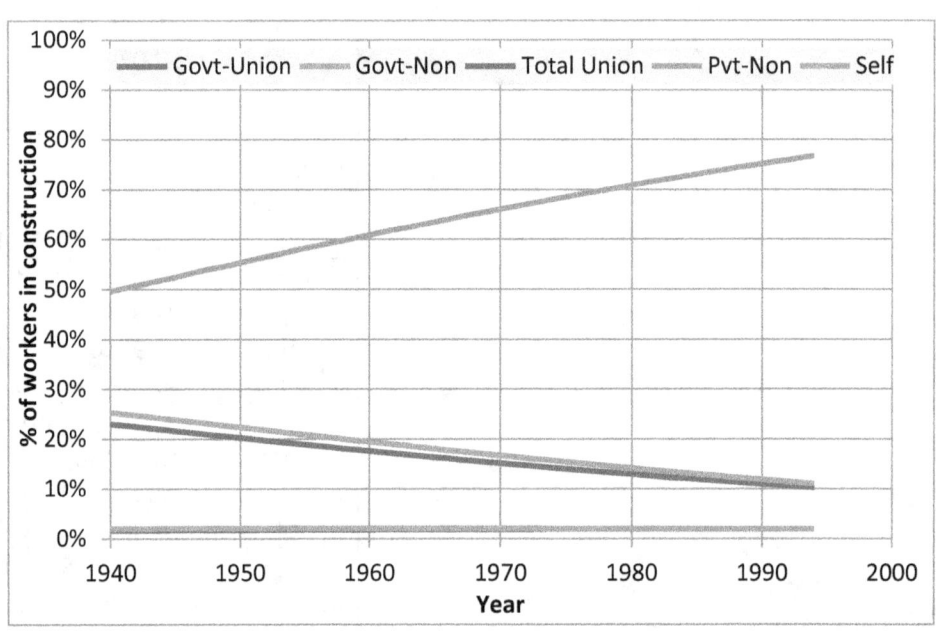

Figure 10: Changes in employer type and union membership with cohort at age 30.

Figure 10 illustrates changes that correlate with cohort. More recent cohorts appear to be less likely to be self-employed. According to the estimates, 30 % of people born in 1960 were likely to be self-employed at age 50, while only an estimated 20 % of those born in 1990 will be self-employed at age 50 (data not shown). Similarly, the decline in union membership appears to be due to younger cohorts choosing not to join rather than to existing members dropping out.

The above regression was rerun while limiting the data to white men with a high-school education only (results not shown). In that regression, there was no significant change in self-employment probability with cohort. Union membership, however, still declined with cohort. That implies that the shift in self-employment probability identified above is a reflection of the changes in the racial and educational composition of the construction work-force (discussed further below). What is not clear is whether the changes in self-employment probability are caused by the demographic changes, or by some other factor that correlates with the demographic changes.

3.2.5. Sex

Table 9 shows the results of the regression for Employment Probability by sex. Men greatly outnumber women in the construction industry. About 7 % of men are in the construction industry, while only about 0.7 % of women are. Results are shown in Figure 11.

Table 9: Coefficients for the estimate of Employment Probability by sex.

	Coef.	Std. Err.	z	Pr
Male				
constant	-2.6485	6.93E-05	-3.80E+04	< 0.001
male ×years	0.003404	8.90E-06	382.45	< 0.001
male × growth	0.163921	0.000089	1842.54	< 0.001
male × unemployment	-0.09696	0.000154	-631.57	< 0.001
Female				
constant	-4.94406	0.00022	-2.30E+04	< 0.001
female ×years	-0.00317	2.89E-05	-109.71	< 0.001
female × growth	0.141863	0.000284	498.98	< 0.001
female × unemployment	-0.19933	0.000498	-400.5	< 0.001

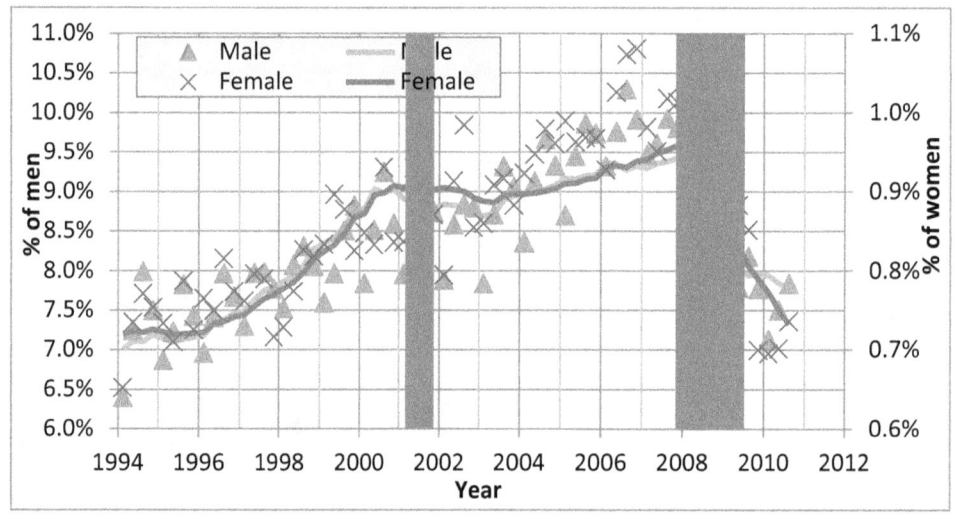

Figure 11: Monthly male and female participation in the construction industry. Left scale is male participation and right scale is female participation.

A Wald test indicates that the likelihood that men are in the industry is increasing at a faster rate than that for women (χ^2 = 5.43, df = 1, p = 0.0198). Over the study period men have increased by 0.4 % while women have increased by 0.04 %.

Women are more susceptible to the business cycle than men. Specifically, the impact of economic growth on female employment is indistinguishable from the impact on men (χ^2 = 1.01, df = 1, p = 0.3161), but the portion of the business cycle reflected in unemployment impacts women harder than men (χ^2 = 8.25, df = 1, p = 0.0041). The difference is probably at least in part due to differences in the occupations they fill in the industry. The Construction labor occupation is dominated by men while the Office and Administrative Support occupations are dominated by women (results not shown). Highest to lowest growth rates reduce the participation of men in the construction industry by about 1.75 % (and women by one tenth of that). Highest to lowest unemployment rates increase the participation of men in the construction industry by about 0.8 %, while it increases women's participation rate by about 1.7 %.

Likelihood of being female given that a worker is in construction is estimated in Table 10.

Table 10: Probability of being female given that a person is employed in construction.

	Coef.	Std. Err.	z	Pr
constant	-2.26426	-0.01565	144.64	< 0.001
years	-0.00754	0.002227	-3.39	0.001
growth	-0.08054	0.066655	-1.21	0.227
unemployment	-0.3644	0.12511	-2.91	0.004

These results are graphed in Figure 12, which represents percent of construction workers who are women. Some 90 % of the people in the construction industry are male. Since the industry is so heavily male, for the remainder of this analysis we will restrict our attention to men.

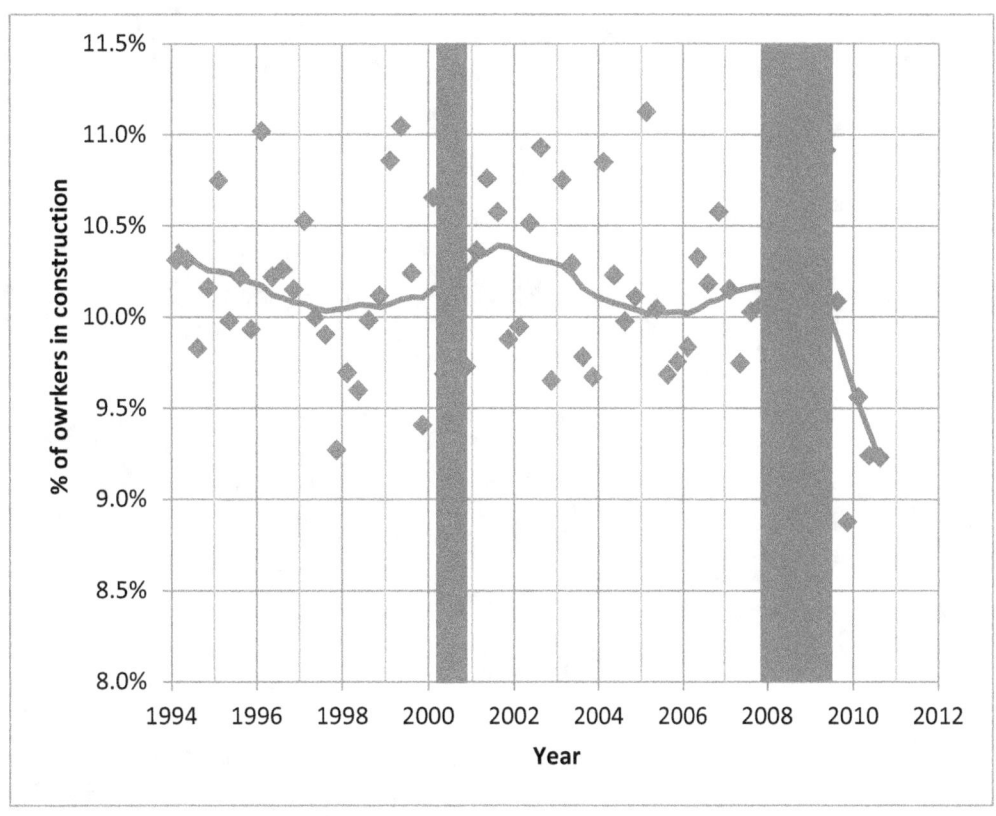

Figure 12: Percent women employed in construction by month.

Additional analysis (not shown) indicates that there are significant differences in female participation in the construction labor force based on education. As educational levels decrease, female participation in the construction labor force also decreases, from about 15 % female for college educated workers down to less than 5 % female for workers without a high-school education.

3.2.6. Age

Results for Employment Probability by age are listed in Table 11, and shown in Figure 13.

Table 11: Coefficients for the estimate of Employment Probability by age.

	Coef.	Std. Err.	z	P>|z|
Men age 16 – 19				
constant	0.02818			
years	-0.02591	0.005047	-5.13	< 0.001
growth	1.69962	0.142559	11.92	< 0.001
unemployment	-0.06796	0.27607	-0.25	0.806
Men age 20 – 24				
constant	0.07512			
years	-0.00486	0.002551	-1.91	0.057
growth	1.05623	0.074985	14.09	< 0.001
unemployment	-0.57991	0.141961	-4.09	< 0.001
Men age 25 – 34				
constant	0.09760			
years	0.00807	0.00145	5.56	< 0.001
growth	0.48462	0.044117	10.98	< 0.001
unemployment	-0.35592	0.082212	-4.33	< 0.001
Men age 35 – 44				
constant	0.09876			
years	0.00894	0.001396	6.4	< 0.001
growth	0.40514	0.042229	9.59	< 0.001
unemployment	-0.16177	0.078807	-2.05	0.04
Men age 45 – 54				
constant	0.07693			
years	0.01961	0.001494	13.13	< 0.001
growth	0.34021	0.046911	7.25	< 0.001
unemployment	-0.12399	0.083663	-1.48	0.138
Men age 55 – 64				
constant	0.05730			
years	0.01022	0.002155	4.74	< 0.001
growth	0.06954	0.068752	1.01	0.312
unemployment	-0.16588	0.121645	-1.36	0.173
Men age 65 and over				
constant	0.00916			
years	0.02910	0.004642	6.27	< 0.001
growth	-0.06576	0.145185	-0.45	0.651
unemployment	-0.85774	0.258386	-3.32	0.001

Probability of being in construction increases with age up to about age 35, peaking at around 10 % of men between 25 and 45. Then it decreases with age till retirement.

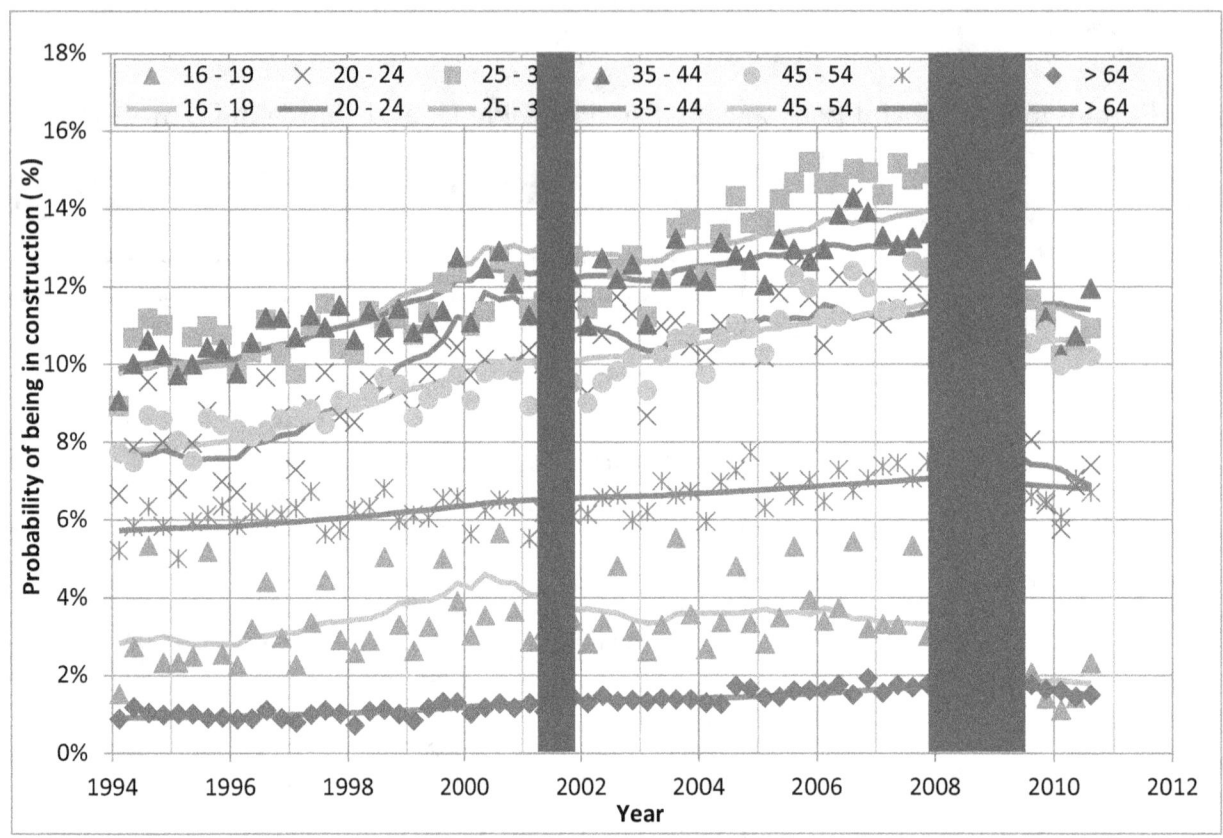

Figure 13: Monthly probability of being in construction given that a man is in a specified age category.

Percentage of men in construction is increasing over time for most age categories (the exceptions being men under the age of 25), with the rate of increase growing at an increasing rate with age. The magnitude of the change over time varies from about 0.5 % for men 65 and over to 1.75 % for men between 45 and 54.

Susceptibility to the business cycle decreases with age. The oldest groups are nearly immune to the business cycle as evidenced by the generally declining coefficients on growth and increasing coefficients on unemployment. Highest to lowest growth rate has essentially no effect on the participation rate of men 65 while it decreases the participation rate of men between 20 and 24 by more than 4 %. Similarly, highest to lowest unemployment rate has essentially no effect on the participation rate of men between 16 and 19, while it increases the participation rate of men between 20 and 24 by about than 1.75 %.

The effect of the broader unemployment rate is the least likely to be significant, but a Likelihood ratio test confirms that as a group they are significant ($\chi^2 = 55.11$, df = 7, $p < 0.0001$). All the effects over age are confirmed with Wald tests.

Estimates of Conditional Probabilities by age are listed in Table 12.

Table 12: Coefficients for the estimate of the Conditional Probability by age.

	Coef.	Std. Err.	z	P>|z|
Age: 16 - 19				
constant	-2.34345	0.033881	-69.17	< 0.001
years	-0.01833	0.004771	-3.84	< 0.001
growth	1.288016	0.151714	8.49	< 0.001
unemployment	0.164418	0.270159	0.61	0.543
Age: 20 - 24				
constant	-1.27325	0.021077	-60.41	< 0.001
years	-0.00009	0.002755	-0.03	0.974
growth	0.428302	0.090166	4.75	< 0.001
unemployment	-0.58331	0.156767	-3.72	< 0.001
Age: 25 - 34				
constant	-0.09583	0.014602	-6.56	< 0.001
years	0.002653	0.001895	1.4	0.161
growth	-0.27502	0.064563	-4.26	< 0.001
unemployment	-0.0958	0.110396	-0.87	0.385
Age: 35 - 44	(base outcome)			
Age: 45 - 54				
constant	-0.54408	0.015887	-34.25	< 0.001
years	0.036427	0.001921	18.96	< 0.001
growth	-0.12569	0.066207	-1.9	0.058
unemployment	-0.13302	0.110659	-1.2	0.229
Age: 55 - 64				
constant	-1.3054	0.020556	-63.51	< 0.001
years	0.047259	0.002355	20.07	< 0.001
growth	-0.65225	0.083179	-7.84	< 0.001
unemployment	-0.06925	0.135519	-0.51	0.609
Age: ≥ 65				
constant	-2.77963	0.038763	-71.71	< 0.001
years	0.044093	0.004313	10.22	< 0.001
growth	-0.77255	0.150636	-5.13	< 0.001
unemployment	-0.57292	0.244189	-2.35	0.019

Marginal effects are listed in Table 13.

Table 13: Marginal Effects on Conditional Probability by age group.

	Age Group						
	16 – 19	20 – 24	25 – 34	35 – 44	45 – 54	55 – 64	≥ 65
Years	-0.00094	-0.00132	-0.00285	-0.00394	0.004972	0.003373	0.000715
Growth	0.040431	0.049635	-0.04271	0.029122	-0.00482	-0.05573	-0.01592
Unemployment	0.008372	-0.04285	0.006997	0.034671	-0.00205	0.005547	-0.01069

Proportion of men over 45 is increasing over time. As expected, younger men are more strongly affected by the business cycle.

Results are shown in Figure 14.

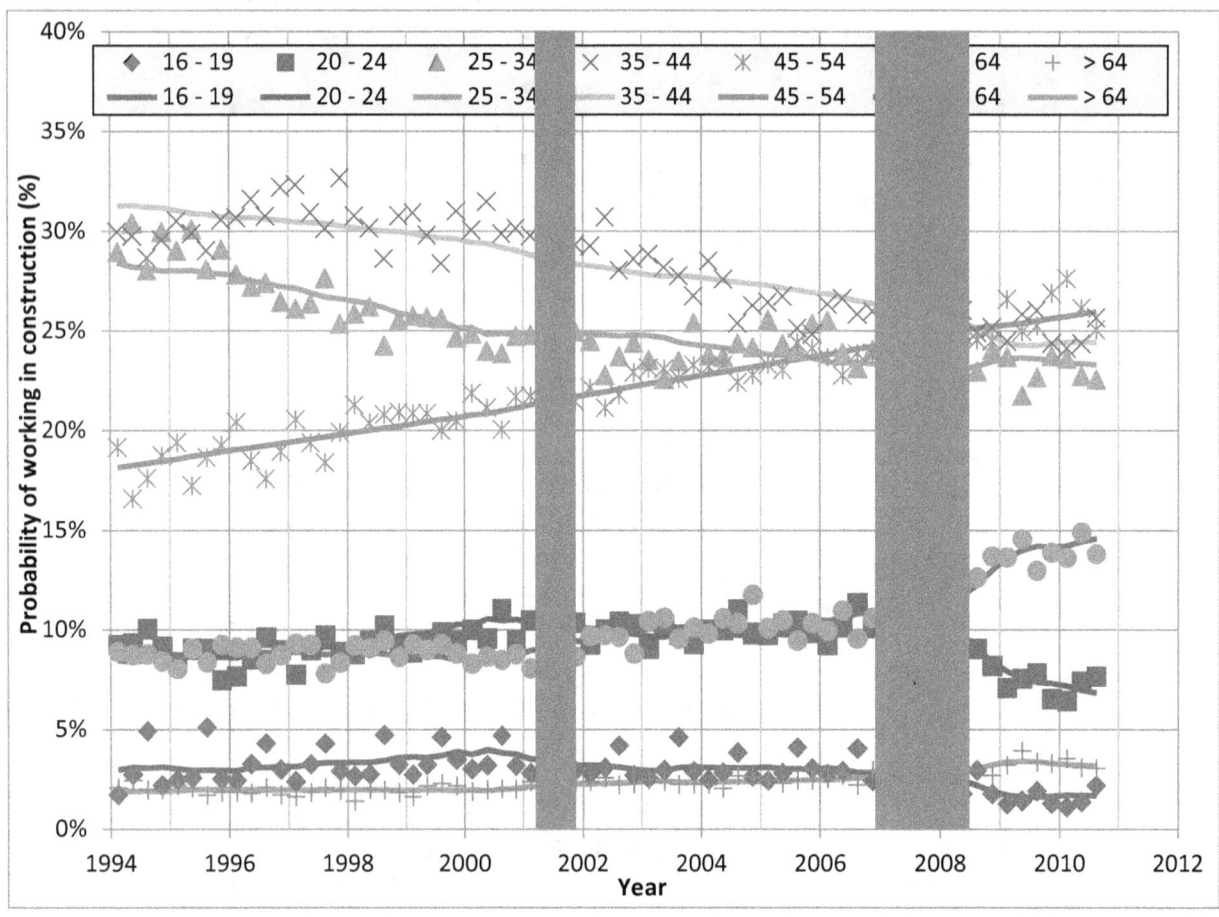

Figure 14: Monthly probability of being in a given age group given that a man is working in construction.

3.2.7. Race and Ethnicity

Coefficients for the estimate of the Employment Probability of the participation of Hispanics, Whites and Blacks in the construction industry are listed in Table 14. For the purposes of this report, people are considered Hispanic if they are reported to be Hispanic by the CPS, regardless of race. All other racial/ethnic categories were dropped since their populations were too small to produce significant results.

Table 14: Coefficients for the estimate of Employment Probability by race/ethnicity.

	Coef.	Std. Err.	z	P>|z|
Black	-3.07451	0.00028	-1.10E+04	< 0.001
Black × years	-0.01737	3.89E-05	-446.12	< 0.001
Black × growth	0.208439	0.000381	547.36	< 0.001
Black × unemployment	-0.06554	0.000676	-96.92	< 0.001
Hispanic	-2.64508	0.000187	-1.40E+04	< 0.001
Hispanic × years	0.034987	1.94E-05	1802.01	< 0.001
Hispanic × growth	0.33116	0.000195	1696.08	< 0.001
Hispanic × unemployment	-0.15741	0.000313	-503.45	< 0.001
White	-2.56821	7.86E-05	-3.30E+04	< 0.001
White × years	-0.00249	1.06E-05	-234.54	< 0.001
White × growth	0.14538	0.000106	1378.27	< 0.001
White × unemployment	-0.06485	0.000185	-350.31	< 0.001

Results are also shown in Figure 15.

Initially, White men were significantly more likely than Hispanic men to be in construction ($\chi^2 = 14.81$, df = 1, p = 0.0001), and Hispanic men were more likely than Black men to be in construction ($\chi^2 = 286$, df = 1, p < 0.0001).

However, the likelihood that Hispanic men are in construction has increased at a much higher rate than that of White (or Black) men ($\chi^2 = 282$, df = 1, p < 0.0001), while the likelihood of Black men in construction is actually decreasing. Hispanic participation rates in construction have increased more than 7 % over the study period, while the White participation rate decreased 0.35 % and the Black participation rate decreased 1.35 %. At present, Hispanic men are more likely to be in construction than White (or Black) men ($\chi^2 = 429$, df = 1, p < 0.0001), with about 12.5 % of Hispanic men in construction compared to 7.6 % of White men and 3.25 % of Black men.

Hispanic men are more susceptible to the business cycle than Black men, while Black and White men are about equally susceptible to the business cycle. A portion of the difference is likely due to differences in occupation within the industry (results not included here indicate that White men are more likely to be in management than the other groups).

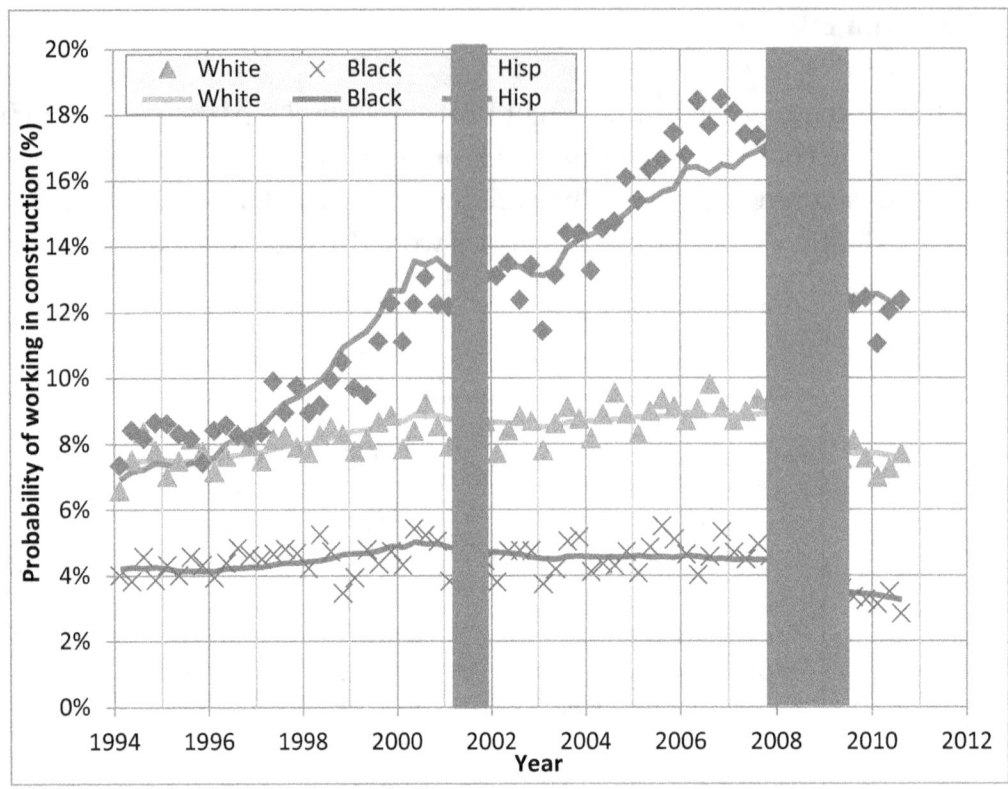

Figure 15: Monthly Employment Probability by race and ethnicity.

Highest to lowest growth rate decreases the participation rate of White men by about 1.5 %, it decreases the participation rate of Black men by about 1.2 %, and it decreases the participation rate of Hispanic men by about 4.9 %. Similarly, highest to lowest unemployment rate increases the participation rate of White men by about 0.6 %, it increases the participation rate of Black men by about 0.3 %, and it increases the participation rate of Hispanic men by about 2 %.

The same information in Figure 15 is presented in a different form in Figure 16. This represents the proportion of men from each race in construction, divided by the proportion of that race in the general population. So values above 1 represent groups that are preferentially concentrated in construction, while numbers less than one represent groups that preferentially avoid construction. As is apparent from the graph, there has been a major change in the employment of Hispanics over the last 20 years, with Hispanics now gravitating toward construction. Meanwhile Blacks gravitate away from it.

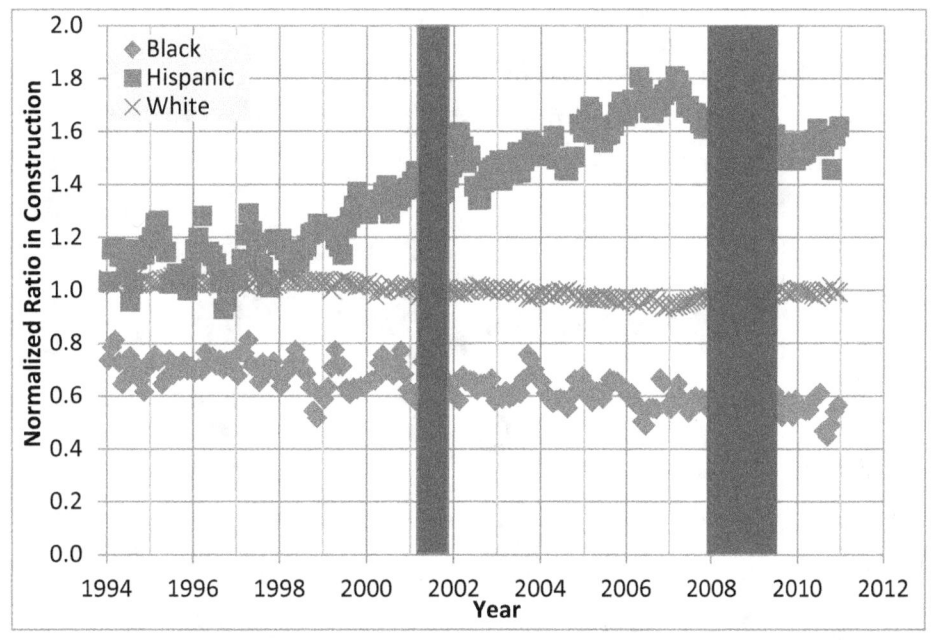

Figure 16: For each race/ethnicity, the fraction of men working in construction divided by the proportion of that race/ethnicity in the population, by month.

Estimates of the Conditional Probability by race and ethnicity are listed in Table 15.

Table 15: Coefficients for the estimate of the Conditional Probability by race and ethnicity

	Coef.	Std. Err.	z	P>|z|
	Black			
constant	-2.77323	0.025509	-108.72	< 0.001
years	-0.01279	0.003813	-3.35	0.001
growth	0.150975	0.113599	1.33	0.184
unemployment	-0.11188	0.215452	-0.52	0.604
	Hispanic			
constant	-2.46572	0.018306	-134.69	< 0.001
years	0.070575	0.002122	33.26	< 0.001
growth	0.880768	0.065718	13.4	< 0.001
unemployment	-0.08542	0.115364	-0.74	0.459
	White (base outcome)			

Marginal effects are listed in Table 16. Results are displayed in Figure 17.

Table 16: Marginal effects on Conditional Probability by race and ethnicity.

	White	Black	Hispanic
Years	-0.0081	-0.0011	0.0092
Growth	-0.1132	0.0004	0.1128
Unemployment	0.0146	-0.0044	-0.0102

Unemployment is jointly indistinguishable from zero ($\chi^2 = 0.75$, df = 2, p = 0.6870). Proportion of Hispanic workers is increasing over time, mostly at the expense of White workers. The business cycle preferentially impacts Hispanic workers.

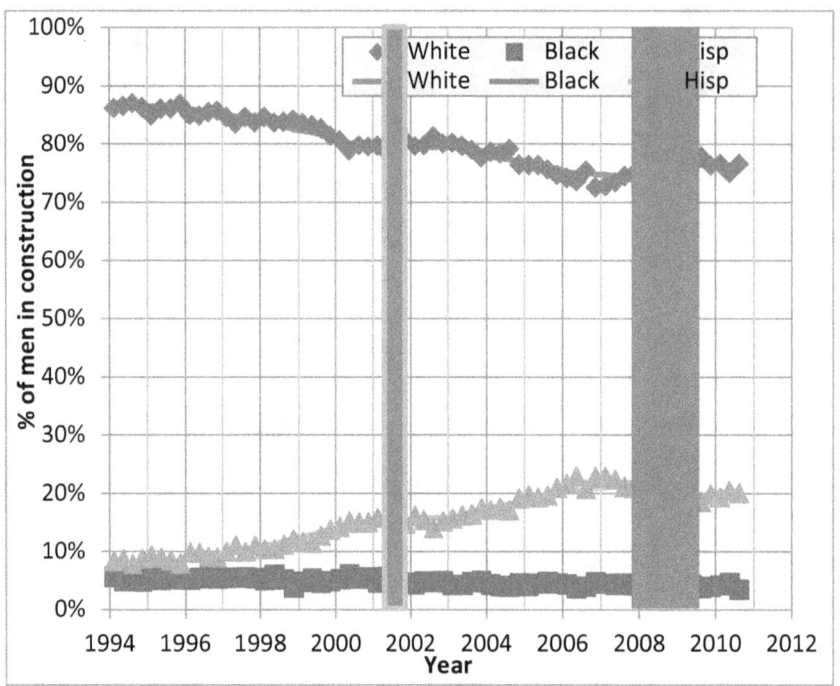

Figure 17: Monthly Conditional Probability by race and ethnicity.

As shown in Figure 18, there is strong regional variation in racial makeup, with the Northeast and Midwest still basically White, the West and increasingly the South becoming strongly Hispanic. The South started with about 10 % Black, but percentage of Black men in construction in the South has dropped dramatically. The West has always had relatively few Black men in construction. The West and the Northeast are the only regions where the proportion of Black men is not declining.

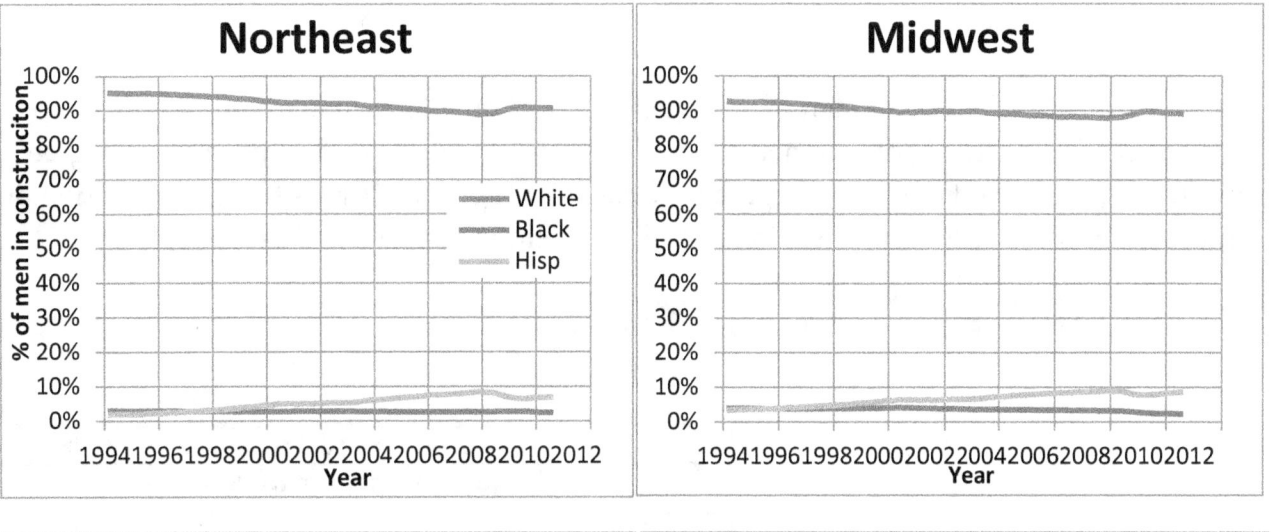

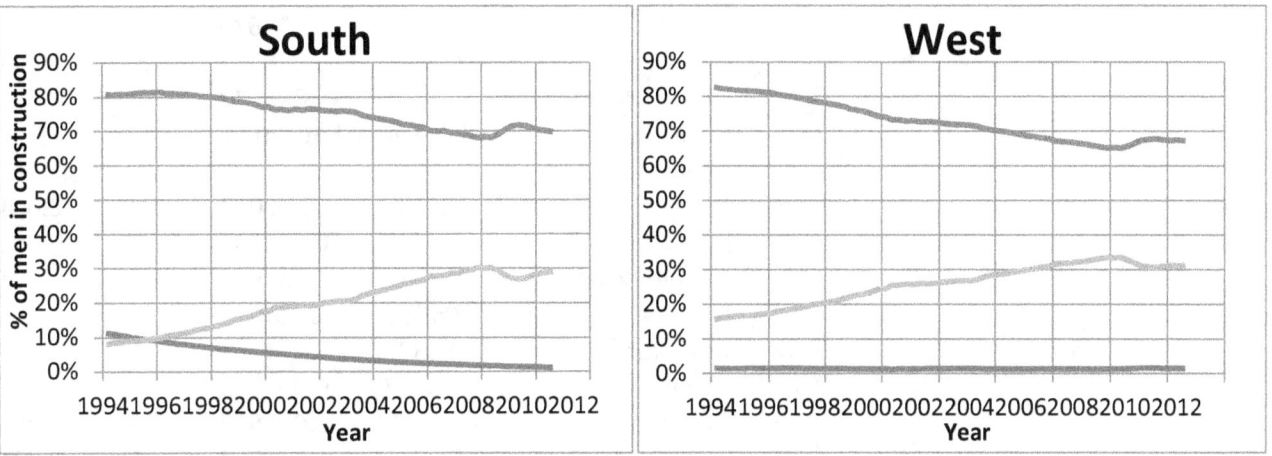

Figure 18: Proportion of men in construction who belong to each race/ethnicity, by region and time.

3.2.8. Education

To analyze educational levels of men in construction, adult men are divided into three groups: those without a high-school education, those with a high-school degree but no college degree, and those with some kind of college degree (associates or higher). Results for the estimation of the Employment Probability by educational level are included in Table 17, and graphed in Figure 19.

Table 17: Coefficients for the estimate of Employment Probability by level of education

	Coef.	Std. Err.	z	P>|z|
No High School Education				
constant	0.0626			
years	0.0202	0.001675	12.04	< 0.001
growth	0.8075	0.050657	15.94	< 0.001
unemployment	-0.1893	0.093932	-2.01	0.044
High School Education				
constant	0.0909			
years	0.0034	0.00092	3.74	< 0.001
growth	0.4446	0.028206	15.76	< 0.001
unemployment	-0.2317	0.052099	-4.45	< 0.001
College Education				
constant	0.0385			
years	0.0127	0.001715	7.43	< 0.001
growth	0.1958	0.053616	3.65	< 0.001
unemployment	-0.3348	0.096367	-3.47	0.001

Initially, those with a high school education were the most likely to be in construction, with men without a high school education second most likely. However, the likelihood of a man without a high school education being in construction has increased at a faster rate than the likelihood for a man with a high school education. Participation rates for men without a high school education in construction have increased about 3 % over the study period, while the participation rate of men with a high school education increased 0.3 % and the participation rate of men with a college education increased about 1 %.

The greater the level of education, the less susceptible a man is to the business cycle. Highest to lowest growth rate decreases the participation rate of men without a high school education by about 3 %, it decreases the participation rate of men with a high school education by about 1.9 %, and it decreases the participation rate of men with some college education by about 0.4 %. Note that the impact of the unemployment rate on each of the groups is jointly indistinguishable from zero according to the Wald test ($\chi^2 = 1.28$, df = 2, p = 0.527).

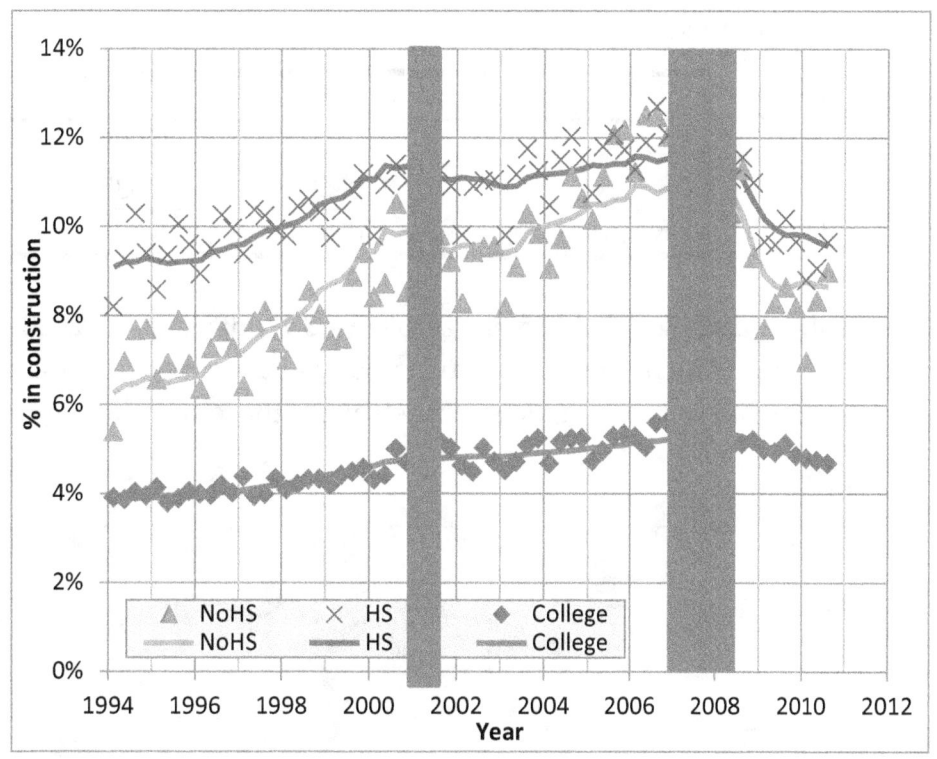

Figure 19: Monthly Employment Probability by level of education.

Estimates for the Conditional Probability by educational level are listed in Table 18.

Table 18: Coefficients for the estimate of the Conditional Probability by educational level.

	Coef.	Std. Err.	z	P>|z|
College				
constant	-1.43391	0.015342	-93.46	< 0.001
years	0.022239	0.002022	11	< 0.001
growth	-0.23127	0.062776	-3.68	< 0.001
unemployment	-0.16721	0.114125	-1.47	0.143
HS (base outcome)				
NoHS				
constant	-1.43391	0.015342	-93.46	< 0.001
years	0.000723	0.00202	0.36	0.72
growth	0.480656	0.061145	7.86	< 0.001
unemployment	0.071576	0.113803	0.63	0.529

Marginal effects are listed in Table 19. Results are displayed in Figure 20.

Table 19: Marginal effects for Conditional Probability by educational level.

	No HS	HS	College
Years	-0.0007	-0.0025	0.0032
Growth	0.0860	-0.0354	-0.0506
Unemployment	0.0175	-0.0092	-0.0268

Unemployment is jointly indistinguishable from zero ($\chi^2 = 3.16$, df = 2, p = 0.2060). Proportion of college educated workers is increasing over time, mostly at the expense of high-school educated workers. The business cycle impacts most strongly those with no high-school education.

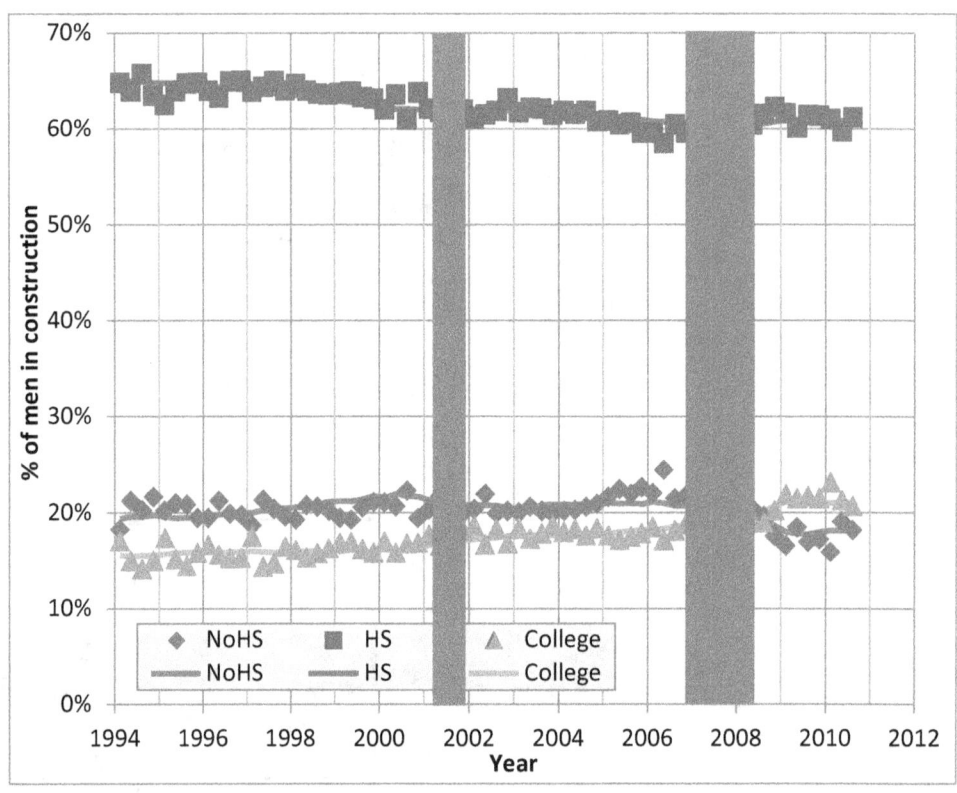

Figure 20: Conditional Probability by educational level.

4. Labor Flows

In order to further evaluate issues regarding the aging of the work force, the number of young people entering the industry, and shifts in the work force toward foreign immigrants, labor flows were estimated. The basic questions answered here is who is entering the work force, who is leaving it, and when.

4.1. Model and Methodology

As before, J represents useful partitions of the population of construction workers. Each member of the partition J is partitioned in turn by cohort. If H is the set of cohorts, then the final partition used here will be $\bar{J} = \{J \times H\}$. Then for any $k \in \bar{J}$, N_t^k is the number of workers in group j in construction at time t, and n_t^k is the number of workers in group j in construction at time t in the CPS sample.

In general, assume

$$n_t^k \sim nb(r_t \cdot m_t^k, q_t^k)$$

Where, "nb" represents the negative binomial distribution, r_t is the sampling rate at time t, and $m_t^k = r_t \cdot N_t^k$. The negative binomial distribution is used to allow for the possibility that there is over-dispersion in the selection of people into a particular partition.

The use of r_t gives equal weight to each observation at a given time. The CPS contains detailed weighting information for each observation, and those weights sometimes differ significantly from individual to individual. The use of r_t ignores that information. The principal reason for doing so here is because many of the groups used in the regression below contain zero individuals. Using individual-specific weights could end up giving those observations a disproportionately large weight. Therefore a month-specific average was used.

The value of N_t^k is assumed to have the form:

$$N_t^k = N_{t-1}^k \cdot \exp\left(\beta_1^k(j) + \beta_2^k \frac{D_t}{D} + \beta_3^k(j)(t) + \text{seasonal terms} \right)$$

As before, this can be formulated as:

$$N_t^k = \exp\left(\ln N_{t-1}^k + \beta_1^k(j) + \beta_2^k(j)\ln\frac{D}{D_0} + \beta_3^k(j)(t) + \text{seasonal terms}\right)$$

Net inflows are assumed to vary by age, regardless of cohort. That is, for some $j \in J$, $h \in H$, and time t, a parameter $\beta_t^{\{j,h\}}$ can be expressed as a function $f_a(j, t-h)$.

To simplify the analysis, it is assumed that relationships for net inflows are constant within age groups T defined as:

Age Group	Ages
1	16 – 19
2	20 – 24
3	25 – 34
4	35 – 44
5	45 – 54
6	55 – 64
7	65 – 70

Define the following (overloaded) functions:

$\tau^j(i)$: The time when group j enters age group i.

$\tau^j(i, t) = \max\{\min\{\tau^j(i), t\}, 0\}$

Note that $\tau^j(8)$ is defined as the time when group j turns 71. Then, since the regression variables are piece-wise constant, the relationship for μ becomes:

$$\mu_t^j = \exp\left\{\ln\mu_0^j + \sum_i \phi_i^j(\tau^j(i+1, t) - \tau^j(i, t)) + \sum_i \psi_i \int_{\tau^j(i, t)}^{\tau^j(i+1, t)} \ln(1 + g_s)\, ds + \sum_i \eta_i \int_{\tau^j(i, t)}^{\tau^j(i+1, t)} ds + \text{seasonal}_t \right\}$$

Here g_t is economic growth in time period t.

For the sake of parsimony, seasonal employment is assumed to follow a sinusoidal pattern. That leads to the following parameterization.

$$\mu_t^j = \exp\left\{\ln\mu_0^j + \sum_i \phi_i^j(\tau^j(i+1, t) - \tau^j(i, t)) + \sum_i \psi_i \int_{\tau^j(i, t)}^{\tau^j(i+1, t)} \ln(1 + g_s)\, ds + \sum_i \eta_i \int_{\tau^j(i, t)}^{\tau^j(i+1, t)} ds \right. \\ \left. + \alpha_s^j \sin 2\pi t + \alpha_c^j \cos 2\pi t \right\}$$

To simplify the analysis below, all people in the sample are assumed to be born on January 1 of their cohort year. Since the CPS is a partially longitudinal dataset, and cohort is defined in the sample as Current Year – Age, that means that some people in the sample are assigned to two different cohorts at different times. Due to the nature of the data set, that will apply to about ¼ of the people in the sample. While that introduces some additional dispersion into the results, the impact on the results is likely to be small.

This analysis implicitly assumes that net inflow/outflow rates are static (after accounting for the business cycle and seasonality). However, as will become clear from the results, that assumption is not true. In particular, during the study period, foreign-born Hispanics have entered construction in large numbers. These dynamic changes should be the subject of further study.

4.2. White Men

Figure 21 shows the number of White men in construction in 1994, by age.

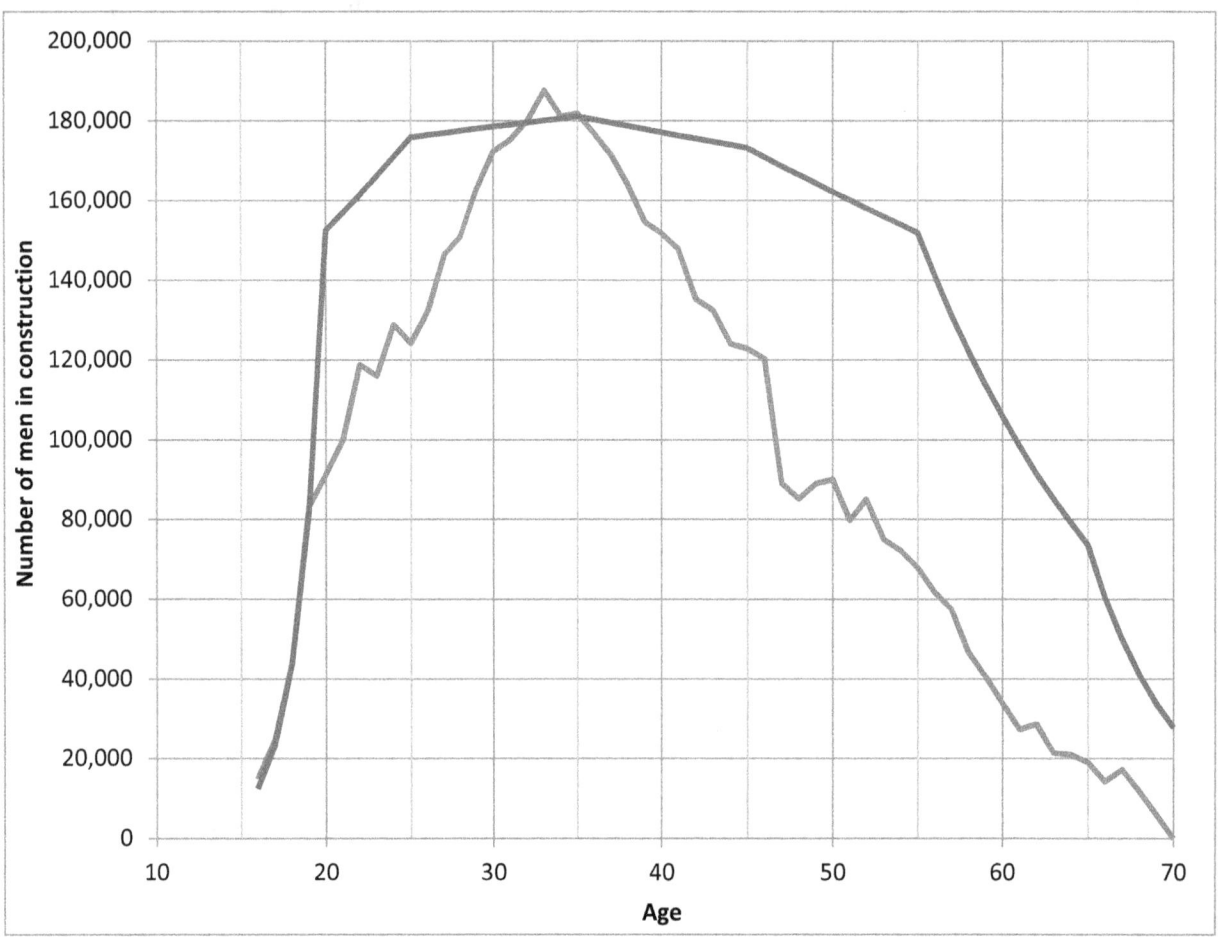

Figure 21: Blue line is the number of White Men in construction in 1994, by age. Red line is the stable number of White men in construction assuming economic growth and unemployment continue at their average rates indefinitely.

Table 20 contains the estimates of the coefficients for net inflows to construction by age group. A stable age distribution of White men in construction assuming economic growth and unemployment continue at their average rates indefinitely is also shown in Figure 21. Fit of the estimate to the data is shown in Figure 22.

Table 20: Coefficients for net flows of White men into construction.

Var	Coef.	Std. Err.	z	P>|z|
Age 16 - 19: t	0.625661	0.012579	49.74	< 0.001
Age 16 - 19: growth	0.280391	0.069935	4.01	< 0.001
Age 16 - 19: unemployment	0.1152849	0.117699	0.98	0.327
Age 16 - 19: sin	-0.1061827	0.020601	-5.15	< 0.001
Age 16 - 19: cos	-0.294515	0.020652	-14.26	< 0.001
Age 20 - 24: t	0.0286478	0.004517	6.34	< 0.001
Age 20 - 24: growth	0.2942997	0.028535	10.31	< 0.001
Age 20 - 24: unemployment	-0.0118243	0.049897	-0.24	0.813
Age 20 - 24: sin	-0.0878365	0.01113	-7.89	< 0.001
Age 20 - 24: cos	-0.0981155	0.01106	-8.87	< 0.001
Age 25 - 34: t	0.002898	0.002194	1.32	0.187
Age 25 - 34: growth	0.124388	0.017583	7.07	< 0.001
Age 25 - 34: unemployment	-0.1096004	0.031504	-3.48	0.001
Age 25 - 34: sin	-0.0487273	0.006955	-7.01	< 0.001
Age 25 - 34: cos	-0.0314539	0.006915	-4.55	< 0.001
Age 35 - 44: t	-0.0044467	0.001952	-2.28	0.023
Age 35 - 44: growth	0.07624	0.015831	4.82	< 0.001
Age 35 - 44: unemployment	-0.0651126	0.02841	-2.29	0.022
Age 35 - 44: sin	-0.0290964	0.006332	-4.59	< 0.001
Age 35 - 44: cos	-0.0291836	0.0063	-4.63	< 0.001
Age 45 - 54: t	-0.0131271	0.001999	-6.57	< 0.001
Age 45 - 54: growth	0.0871276	0.016806	5.18	< 0.001
Age 45 - 54: unemployment	-0.0468046	0.028574	-1.64	0.101
Age 45 - 54: sin	-0.0578177	0.006808	-8.49	< 0.001
Age 45 - 54: cos	-0.0103273	0.006767	-1.53	0.127
Age 55 - 64: t	-0.0723524	0.00283	-25.56	< 0.001
Age 55 - 64: growth	-0.051779	0.023582	-2.2	0.028
Age 55 - 64: unemployment	-0.0686209	0.038646	-1.78	0.076
Age 55 - 64: sin	-0.0590978	0.009733	-6.07	< 0.001
Age 55 - 64: cos	-0.0317859	0.00966	-3.29	0.001
Age > 64: t	-0.1944142	0.008944	-21.74	< 0.001
Age > 64: growth	0.0001219	0.05949	0	0.998
Age > 64: unemployment	-0.2762774	0.094223	-2.93	0.003
Age > 64: sin	-0.1528037	0.023332	-6.55	< 0.001
Age > 64: cos	-0.0395778	0.023005	-1.72	0.085

In general, as noted before, older worker are less susceptible to the business cycle than younger workers. Above the age of 55, the impact of the business cycle is not significantly different from zero.

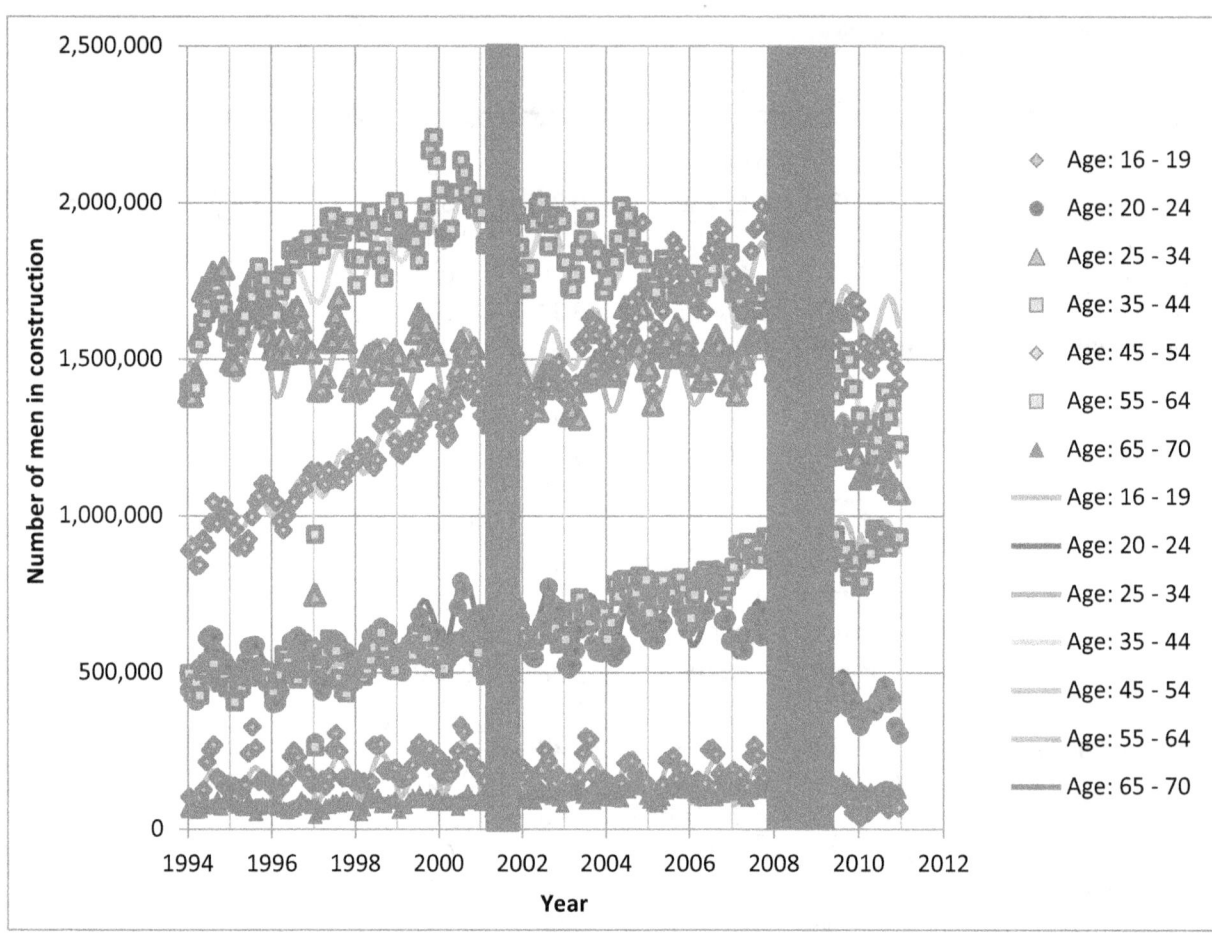

Figure 22: Fit of estimates to data for White men in construction by age group.

Net inflow totals are shown in Figure 23. In the average year, 75 % of the White men who enter construction do so before the age of 21. Total net inflow across all age groups is positive in all but five years (data not shown).

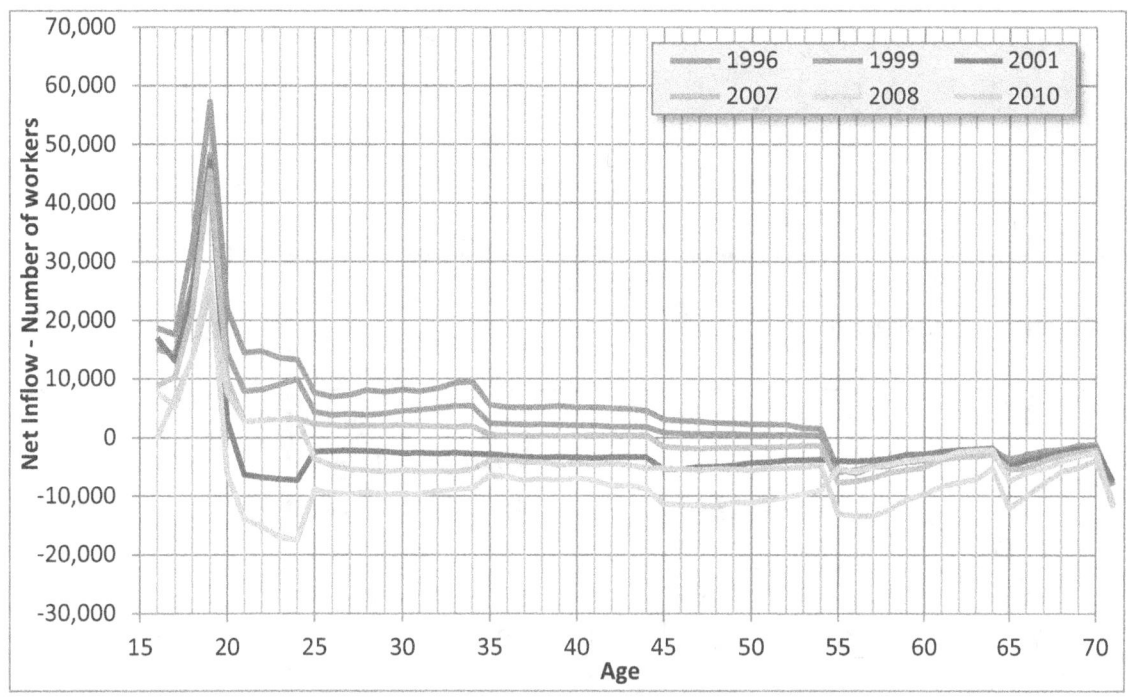

Figure 23: Net inflow to construction for White men, by age for selected years.

Net rate of inflow by age and rate of economic growth is shown in Table 21. Note that this underestimates the variation due to the business cycle, since it does not take into account the general unemployment rate. Inflow rates are very large for the 16 – 19 year old group because the initial number of white men in construction on turning 16 is so small.

Table 21: Net rate of inflow to construction by age and rate of economic growth. Colored cells indicate ages and circumstances where men are on net leaving construction.

Age	Deviation of Economic Growth from Average				
	-2 %	-1 %	0 %	1 %	2 %
16 – 19	74.66 %	80.73 %	86.95 %	93.31 %	99.83 %
20 – 24	-4.18 %	-0.68 %	2.91 %	6.59 %	10.36 %
25 – 35	-2.69 %	-1.20 %	0.29 %	1.79 %	3.30 %
35 – 44	-2.27 %	-1.35 %	-0.44 %	0.47 %	1.38 %
45 – 54	-3.37 %	-2.34 %	-1.30 %	-0.27 %	0.76 %
55 – 64	-5.80 %	-6.40 %	-6.98 %	-7.55 %	-8.12 %
65 – 70	-17.67 %	-17.67 %	-17.67 %	-17.67 %	-17.67 %

Impact of seasonality is shown in Table 22. As expected, the impact of seasonality decreases as workers get older, reaching a minimum at around age 40. Then seasonality increases slowly. The relatively large increase in seasonality for workers over the age of 65 probably reflects semi-retired workers returning to the work force when demand is high. The date when number of workers is highest is concentrated in late summer. Interestingly enough, date of peak employment is progressively later the older the worker gets. A Wald test on date of maximum employment indicates that the dates are not all equal across age groups ($\chi^2 = 7.80$, df = 6, p = 0.01 %).

Table 22: Impact of seasonality. Seasonal Coefficient is the absolute magnitude of the seasonal variation. Wald Test represents the probability that the seasonal variation is greater than zero. Percent change is calculated based on the highest estimated seasonal employment compared to the lowest estimated seasonal employment.

Age	Seasonal Coefficient	% change: lowest to highest	Wald test			Max Date
			χ^2	df	p	
16 – 19	0.3131	87.04 %	56.23	1	< 0.01 %	20-Jul
20 – 24	0.1317	30.13 %	35.32	1	< 0.01 %	11-Aug
25 – 35	0.0580	12.30 %	17.44	1	< 0.01 %	27-Aug
35 – 44	0.0412	8.59 %	10.64	1	0.11 %	15-Aug
45 – 54	0.0587	12.46 %	18.64	1	< 0.01 %	19-Sep
55 – 64	0.0671	14.36 %	11.97	1	0.05 %	1-Sep
65 – 70	0.1578	37.12 %	11.58	1	0.07 %	15-Sep

4.3. Black Men

Figure 24 shows the number of Black men in construction in 1994, by age.

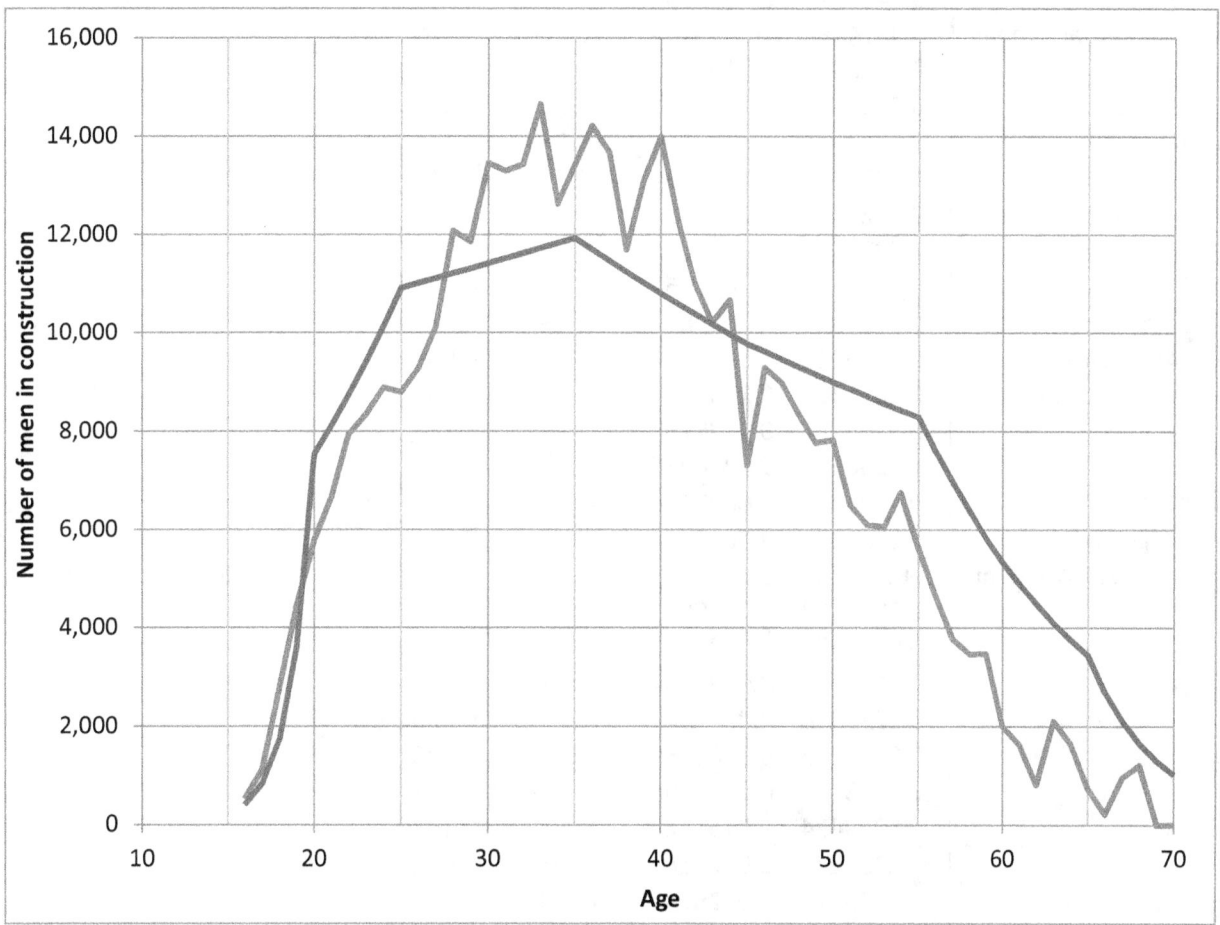

Figure 24: Blue line is the number of Black Men in construction in 1994, by age. Red line is the stable number of Black men in construction assuming economic growth and unemployment continue at their average rates indefinitely.

Table 23 contains the estimates of the coefficients for net inflows to construction by age group. A stable age distribution of Black men in construction assuming economic growth and unemployment continue at their average rates indefinitely is shown in Figure 24. Fit of the estimate to the data is shown in Figure 25.

Table 23: Coefficients for net flows of Black men into construction.

Var	Coef.	Std. Err.	z	P>|z|
Age 16 - 19: t	0.7353515	0.070693	10.4	< 0.001
Age 16 - 19: growth	-0.0413578	0.393725	-0.11	0.916
Age 16 - 19: unemployment	-0.4334484	0.65523	-0.66	0.508
Age 16 - 19: sin	-0.0681683	0.10814	-0.63	0.528
Age 16 - 19: cos	-0.3085816	0.108717	-2.84	0.005
Age 20 - 24: t	0.0742974	0.02003	3.71	< 0.001
Age 20 - 24: growth	0.1477798	0.126993	1.16	0.245
Age 20 - 24: unemployment	-0.3426494	0.227113	-1.51	0.131
Age 20 - 24: sin	-0.0020518	0.049001	-0.04	0.967
Age 20 - 24: cos	-0.09982	0.048746	-2.05	0.041
Age 25 - 34: t	0.0088826	0.008981	0.99	0.323
Age 25 - 34: growth	0.2169711	0.070975	3.06	0.002
Age 25 - 34: unemployment	-0.1671427	0.129634	-1.29	0.197
Age 25 - 34: sin	-0.0396778	0.027576	-1.44	0.15
Age 25 - 34: cos	0.0286096	0.027433	1.04	0.297
Age 35 - 44: t	-0.0198847	0.007726	-2.57	0.01
Age 35 - 44: growth	0.2005771	0.061379	3.27	0.001
Age 35 - 44: unemployment	-0.175498	0.112466	-1.56	0.119
Age 35 - 44: sin	-0.0870369	0.023908	-3.64	< 0.001
Age 35 - 44: cos	-0.0732338	0.023798	-3.08	0.002
Age 45 - 54: t	-0.0165047	0.007478	-2.21	0.027
Age 45 - 54: growth	0.1726808	0.063152	2.73	0.006
Age 45 - 54: unemployment	0.0378783	0.108248	0.35	0.726
Age 45 - 54: sin	0.0096741	0.025269	0.38	0.702
Age 45 - 54: cos	-0.040474	0.025142	-1.61	0.107
Age 55 - 64: t	-0.087921	0.011114	-7.91	< 0.001
Age 55 - 64: growth	0.178611	0.092684	1.93	0.054
Age 55 - 64: unemployment	-0.0033049	0.153552	-0.02	0.983
Age 55 - 64: sin	-0.1034498	0.036654	-2.82	0.005
Age 55 - 64: cos	-0.0298301	0.03639	-0.82	0.412
Age > 64: t	-0.2452788	0.037595	-6.52	< 0.001
Age > 64: growth	-0.1290224	0.242949	-0.53	0.595
Age > 64: unemployment	-0.534806	0.389604	-1.37	0.17
Age > 64: sin	-0.2398704	0.09528	-2.52	0.012
Age > 64: cos	-0.0393219	0.093518	-0.42	0.674

As usual, older worker are less susceptible to the business cycle than younger workers. Above the age of 55, the impact of the business cycle is not significantly different from zero.

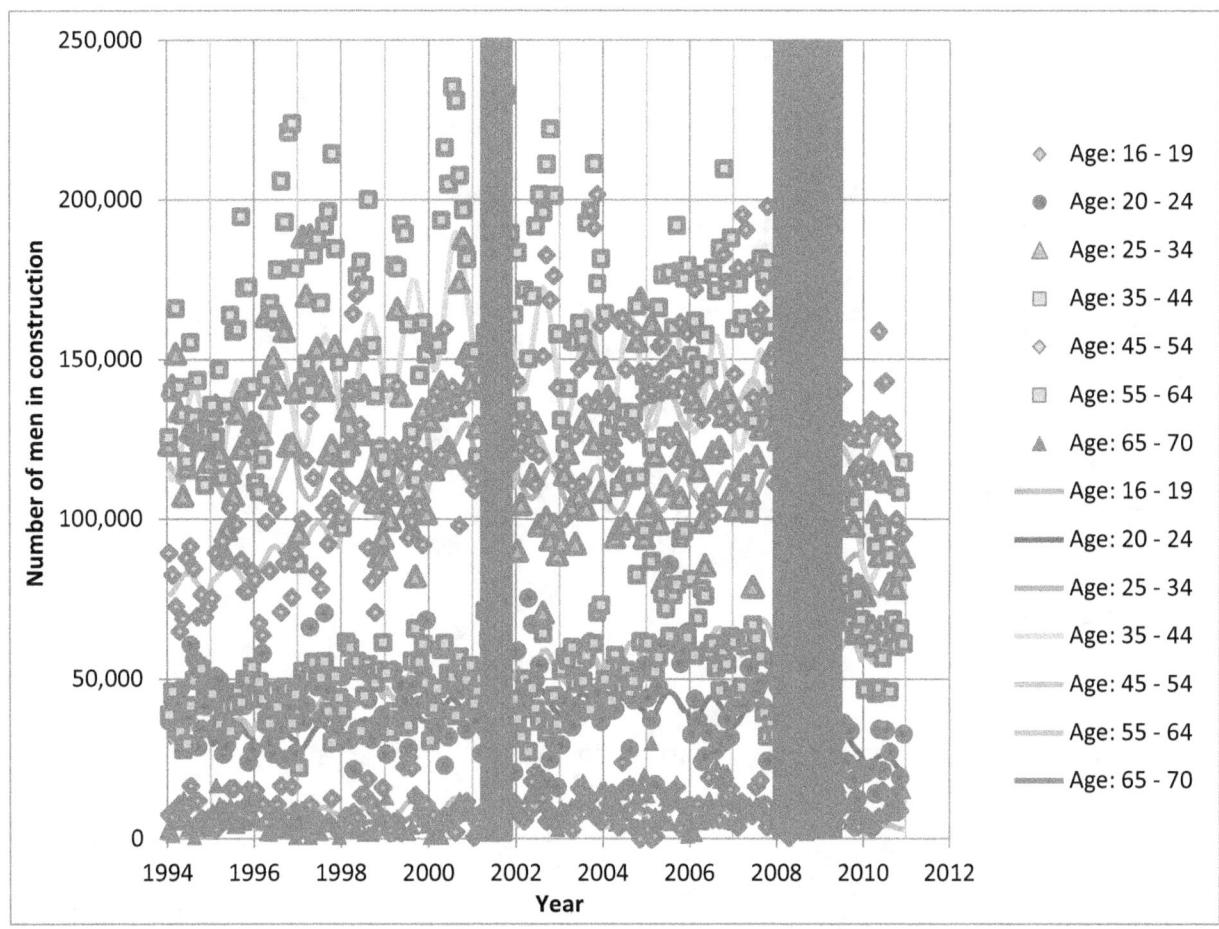

Figure 25: Fit of estimates to data for Black men in construction by age group, by month.

Net inflow totals are shown in Figure 26. In the average year, 75 % of the Black men who enter construction do so before the age of 24. Total net inflow across all age groups is positive in all but six years (data not shown).

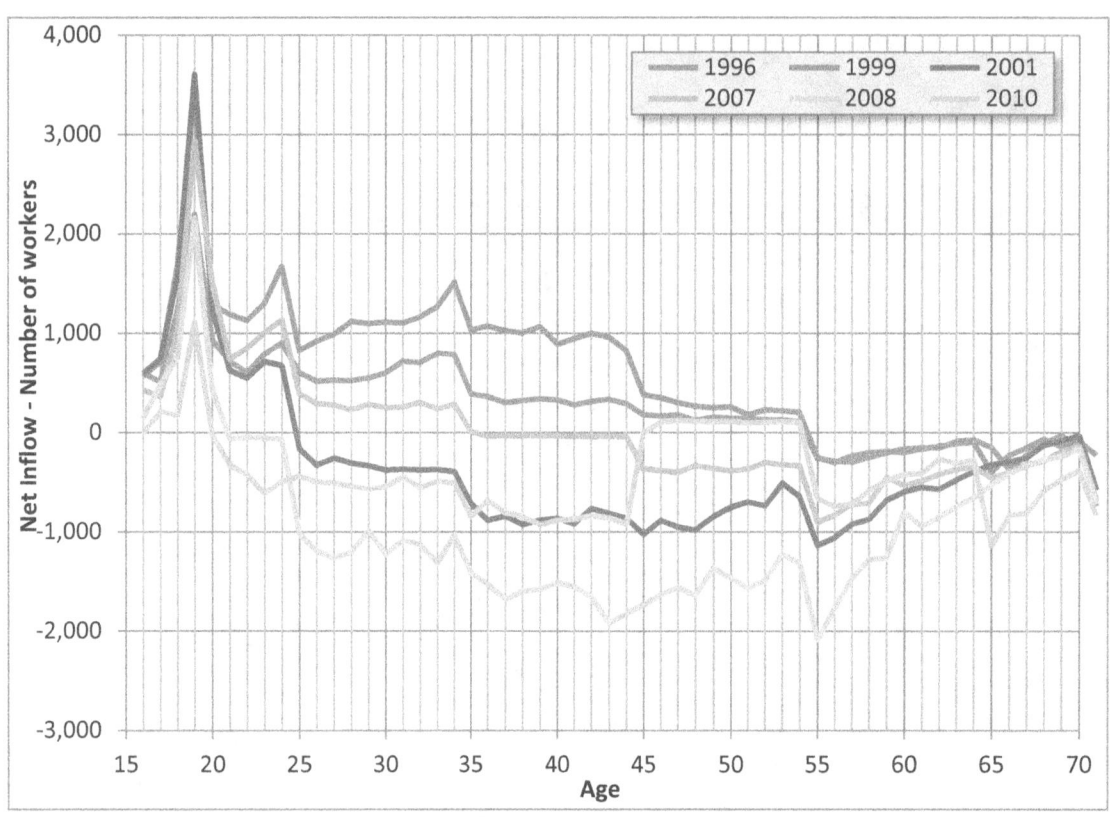

Figure 26: Net inflow to construction for Black men, by age for selected years.

Net rate of inflow by age and rate of economic growth is shown in Table 24. As before, this underestimates the variation due to the business cycle, since it does not take into account the general unemployment rate.

Table 24: Net rate of inflow to construction by age and rate of economic growth. Colored cells indicate ages and circumstances where men are on net leaving construction.

| | Deviation of Economic Growth from Average | | | | |
Age	-2 %	-1 %	0 %	1 %	2 %
16 – 19	110.72 %	109.66 %	108.62 %	107.59 %	106.58 %
20 – 24	3.92 %	5.81 %	7.71 %	9.63 %	11.56 %
25 – 35	-4.28 %	-1.71 %	0.89 %	3.54 %	6.23 %
35 – 44	-6.62 %	-4.31 %	-1.97 %	0.41 %	2.82 %
45 – 54	-5.67 %	-3.66 %	-1.64 %	0.41 %	2.48 %
55 – 64	-12.30 %	-10.37 %	-8.42 %	-6.44 %	-4.45 %
65 – 70	-19.27 %	-20.52 %	-21.75 %	-22.95 %	-24.11 %

Impact of seasonality is shown in Table 25. As expected, the impact of seasonality decreases as workers get older, reaching a minimum at around age 50, then seasonality increases again. However, seasonality is not significantly different from zero for any age group except 35 to 44 year-olds. The date when number of workers is highest is concentrated in summer. A Wald test on date of maximum employment indicates that the dates are not significantly different across age groups ($X^2 = 5.88$, df = 6, p = 43.69 %).

Table 25: Impact of seasonality. Seasonal Coefficient is the absolute magnitude of the seasonal variation. Wald Test represents the probability that the seasonal variation is greater than zero.

Age	Seasonal Coefficient	% change: lowest to highest	Wald test χ^2	df	p	Max Date
16 – 19	0.3160	88.15 %	2.070	1	15.03 %	13-Jul
20 – 24	0.0998	22.10 %	1.050	1	30.58 %	1-Jul
25 – 35	0.0489	10.28 %	0.790	1	37.38 %	5-Nov
35 – 44	0.1137	25.55 %	5.700	1	1.70 %	20-Aug
45 – 54	0.0416	8.68 %	0.680	1	40.85 %	16-Jun
55 – 64	0.1077	24.03 %	2.170	1	14.07 %	13-Sep
65 – 70	0.2431	62.60 %	1.650	1	19.93 %	20-Sep

4.4. U.S. Born Hispanic Men

Figure 27 shows the number of U.S. born Hispanic men in construction in 1994, by age.

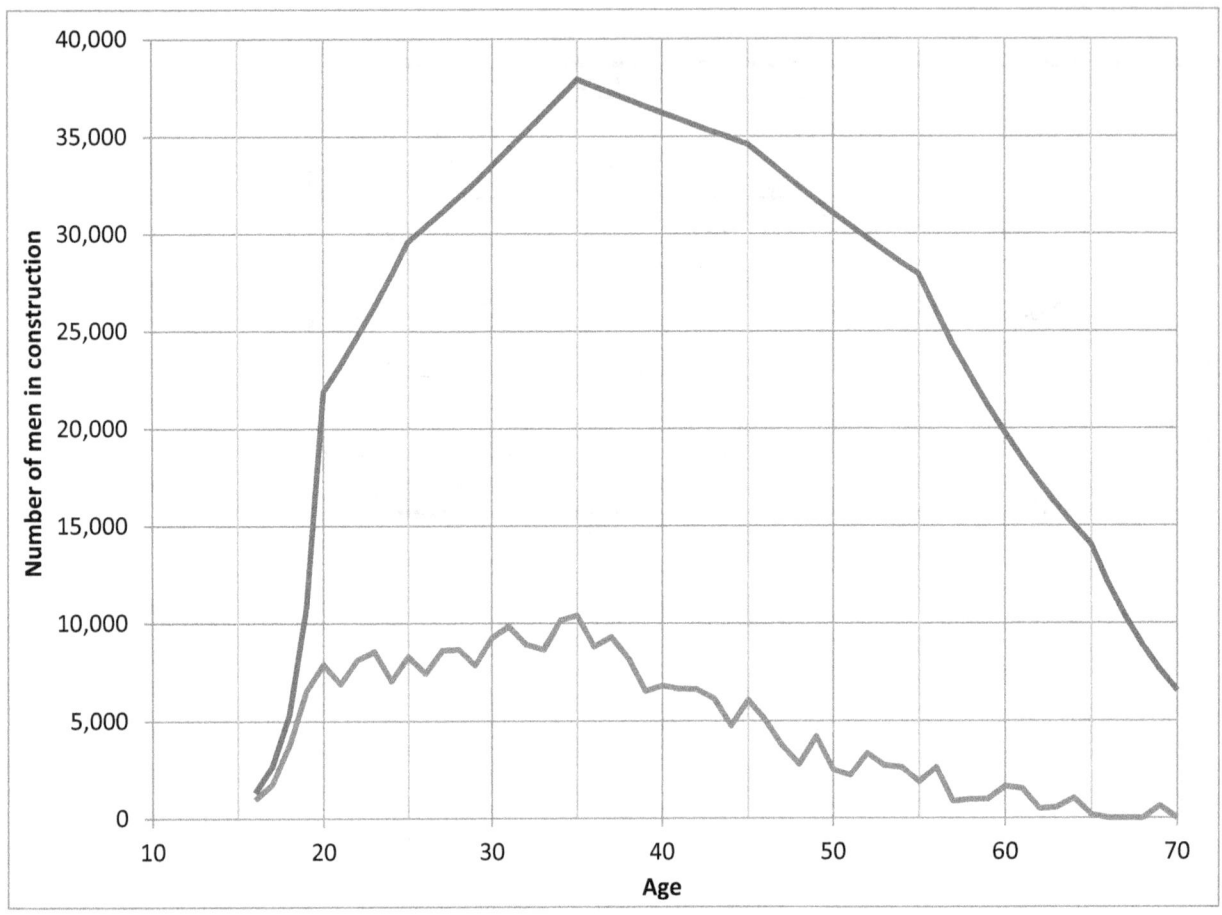

Figure 27: Blue line is the number of US Born Hispanic Men in construction in 1994, by age. Red line is the stable number of US Born Hispanic men in construction assuming economic growth and unemployment continue at their average rates indefinitely.

Table 26 contains the estimates of the coefficients for net inflows to construction by age group. A stable age distribution of U.S. born Hispanic men in construction assuming economic growth and unemployment continue at their average rates indefinitely is shown in Figure 27. The "stable age distribution" is well above the 1994 age distribution because of the rapid influx of Hispanics into the construction industry. Fit of the estimate to the data is shown in Figure 28.

Table 26: Coefficients for net flows of US Born Hispanic men into construction.

Var	Coef.	Std. Err.	z	P>|z|
Age 16 - 19: t	0.701	0.0410	17.120	< 0.01 %
Age 16 - 19: growth	0.573	0.2194	2.610	0.90 %
Age 16 - 19: unemployment	0.201	0.3458	0.580	56.20 %
Age 16 - 19: sin	0.113	0.0665	1.700	8.90 %
Age 16 - 19: cos	-0.078	0.0658	-1.190	23.50 %
Age 20 - 24: t	0.060	0.0133	4.530	< 0.01 %
Age 20 - 24: growth	0.227	0.0854	2.660	0.80 %
Age 20 - 24: unemployment	-0.273	0.1429	-1.910	5.60 %
Age 20 - 24: sin	-0.068	0.0330	-2.060	3.90 %
Age 20 - 24: cos	-0.027	0.0328	-0.840	40.30 %
Age 25 - 34: t	0.025	0.0070	3.520	< 0.01 %
Age 25 - 34: growth	0.127	0.0580	2.190	2.90 %
Age 25 - 34: unemployment	-0.272	0.0987	-2.760	0.60 %
Age 25 - 34: sin	-0.021	0.0230	-0.900	37.00 %
Age 25 - 34: cos	-0.025	0.0229	-1.110	26.90 %
Age 35 - 44: t	-0.009	0.0077	-1.190	23.40 %
Age 35 - 44: growth	0.064	0.0634	1.020	30.90 %
Age 35 - 44: unemployment	-0.230	0.1095	-2.100	3.60 %
Age 35 - 44: sin	-0.004	0.0257	-0.140	88.70 %
Age 35 - 44: cos	-0.073	0.0256	-2.850	0.40 %
Age 45 - 54: t	-0.021	0.0096	-2.250	2.50 %
Age 45 - 54: growth	0.181	0.0808	2.230	2.50 %
Age 45 - 54: unemployment	-0.152	0.1347	-1.130	25.80 %
Age 45 - 54: sin	-0.088	0.0327	-2.690	0.70 %
Age 45 - 54: cos	-0.005	0.0324	-0.140	88.60 %
Age 55 - 64: t	-0.069	0.0157	-4.370	< 0.01 %
Age 55 - 64: growth	0.278	0.1313	2.120	3.40 %
Age 55 - 64: unemployment	-0.039	0.2140	-0.180	85.70 %
Age 55 - 64: sin	0.042	0.0531	0.780	43.30 %
Age 55 - 64: cos	0.063	0.0528	1.190	23.50 %
Age > 64: t	-0.153	0.0493	-3.110	0.20 %
Age > 64: growth	-0.021	0.3359	-0.060	95.10 %
Age > 64: unemployment	0.105	0.5129	0.210	83.70 %
Age > 64: sin	0.045	0.1344	0.330	73.90 %
Age > 64: cos	-0.171	0.1345	-1.270	20.30 %

As usual, older worker are less susceptible to the business cycle than younger workers. Above the age of 55, the impact of the business cycle is not significantly different from zero.

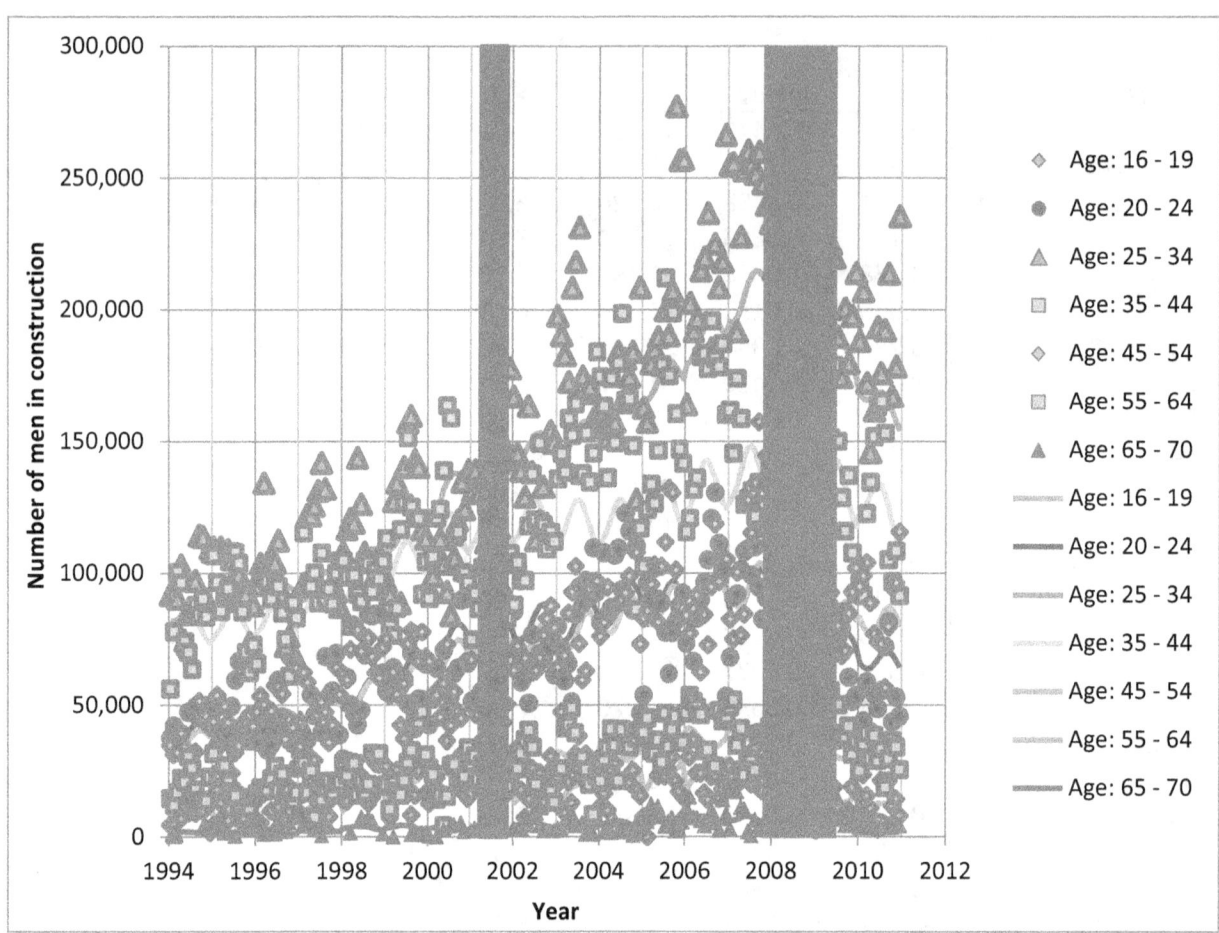

Figure 28: Fit of estimates to data for US Born Hispanic men in construction by age group.

Net inflow totals are shown in Figure 29. In the average year, 75 % of the U. S. born Hispanic men who enter construction do so before the age of 25. Total net inflow across all age groups is positive in all but three years (data not shown).

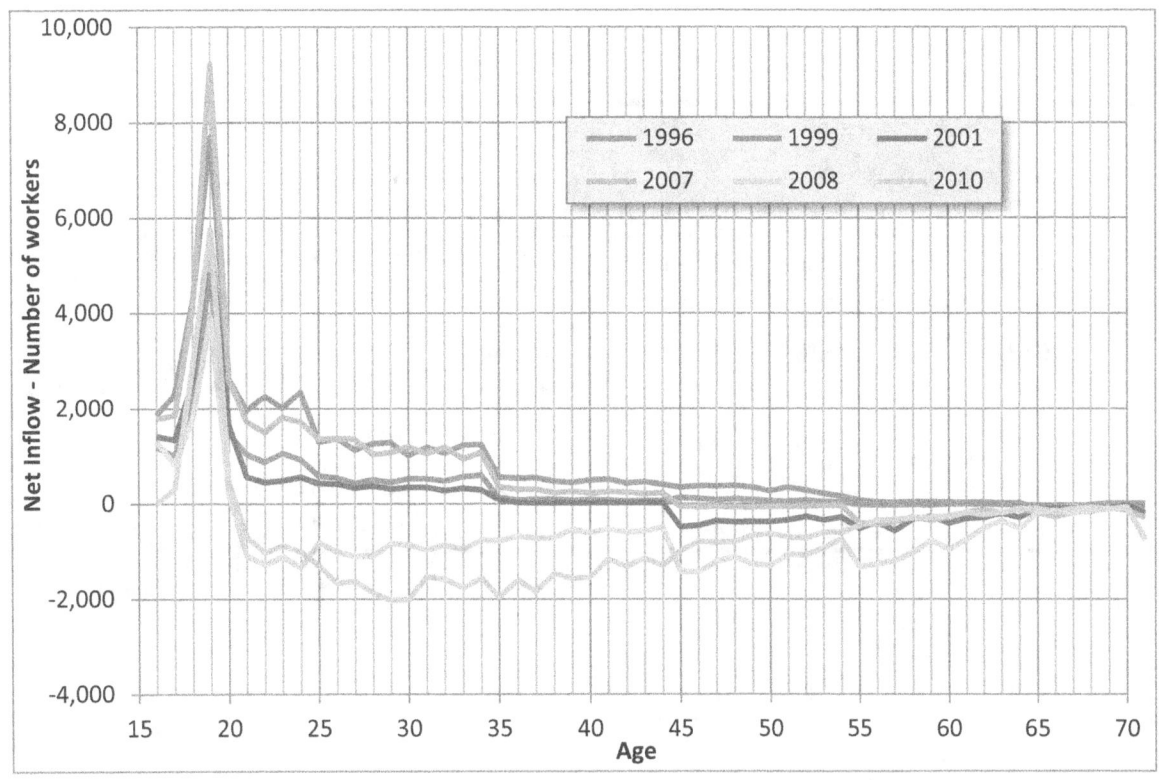

Figure 29: Net inflow to construction for U.S. born Hispanic men, by age for selected years.

Net rate of inflow by age and rate of economic growth is shown in Table 27. As before, this underestimates the variation due to the business cycle, since it does not take into account the general unemployment rate. Here, susceptibility to the business cycle reaches a minimum at around 40 (and above 65).

Table 27: Net rate of inflow to construction by age and rate of economic growth. Colored cells indicate ages and circumstances where men are on net leaving construction.

Age	Deviation of Economic Growth from Average				
	-2 %	-1 %	0 %	1 %	2 %
16 – 19	75.51 %	88.20 %	101.66 %	115.94 %	131.07 %
20 – 24	0.52 %	3.34 %	6.20 %	9.12 %	12.08 %
25 – 35	-0.59 %	0.96 %	2.51 %	4.08 %	5.65 %
35 – 44	-2.45 %	-1.68 %	-0.91 %	-0.15 %	0.62 %
45 – 54	-6.32 %	-4.23 %	-2.12 %	0.01 %	2.17 %
55 – 64	-12.74 %	-9.73 %	-6.65 %	-3.49 %	-0.26 %
65 – 70	-13.76 %	-13.98 %	-14.19 %	-14.40 %	-14.61 %

Impact of seasonality is shown in Table 28. Seasonality is not significantly different from zero for U.S.-born Hispanic men. The date when number of workers is highest is concentrated in summer. A Wald test on date of maximum employment indicates that the dates are not significantly different across age groups ($\chi^2 = 2.48$, df = 6, p = 87.09 %).

Table 28: Impact of seasonality. Seasonal Coefficient is the absolute magnitude of the seasonal variation. Wald Test represents the probability that the seasonal variation is greater than zero.

Age	Seasonal Coefficient	% change: lowest to highest	Wald test χ^2	df	p	Max Date
16 – 19	0.1375	31.66 %	1.170	1	28.04 %	5-May
20 – 24	0.0734	15.82 %	1.230	1	26.72 %	7-Sep
25 – 35	0.0327	6.75 %	0.500	1	47.84 %	9-Aug
35 – 44	0.0730	15.71 %	2.030	1	15.39 %	3-Jul
45 – 54	0.0879	19.21 %	1.810	1	17.85 %	26-Sep
55 – 64	0.0752	16.24 %	0.510	1	47.60 %	3-Feb
65 – 70	0.1769	42.46 %	0.430	1	51.19 %	15-Jun

4.5.   Foreign Born Hispanic Men

Figure 30 shows the number of foreign-born Hispanic men in construction in 1994, by age.

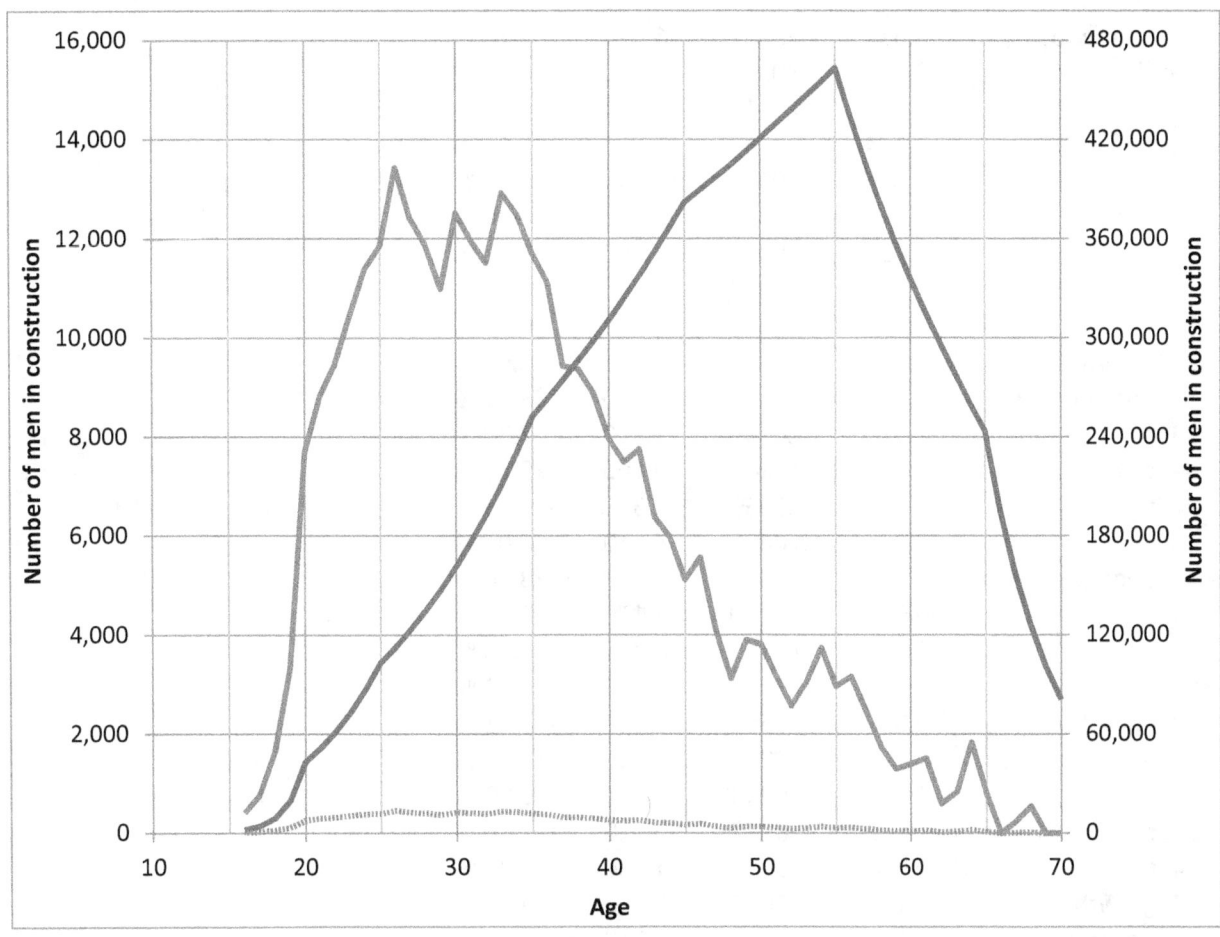

Figure 30: Blue lines both represent the number of foreign-born Hispanic Men in construction in 1994, by age. Red line is the stable number in construction assuming economic growth and unemployment continue at their average rates indefinitely. Note that the solid blue line is plotted against the left axis, while the red line (and dashed blue line for scale) is plotted against the right axis.

Table 29 contains the estimates of the coefficients for net inflows to construction by age group. A stable age distribution of U.S. born Hispanic men in construction assuming economic growth and unemployment continue at their average rates indefinitely is shown in Figure 30. Fit of the estimate to the data is shown in Figure 31.

Table 29: Coefficients for net flows of US Born Hispanic men into construction.

Var	Coef.	Std. Err.	z	P>|z|
Age 16 - 19: t	0.795	0.0336	23.670	< 0.01 %
Age 16 - 19: growth	0.808	0.1750	4.620	< 0.01 %
Age 16 - 19: unemployment	0.519	0.2934	1.770	7.70 %
Age 16 - 19: sin	0.285	0.0522	5.450	< 0.01 %
Age 16 - 19: cos	-0.068	0.0513	-1.320	18.60 %
Age 20 - 24: t	0.174	0.0098	17.720	< 0.01 %
Age 20 - 24: growth	0.646	0.0627	10.310	< 0.01 %
Age 20 - 24: unemployment	-0.343	0.1042	-3.290	0.10 %
Age 20 - 24: sin	0.055	0.0235	2.350	1.90 %
Age 20 - 24: cos	-0.069	0.0233	-2.950	0.30 %
Age 25 - 34: t	0.089	0.0046	19.540	< 0.01 %
Age 25 - 34: growth	0.542	0.0390	13.900	< 0.01 %
Age 25 - 34: unemployment	-0.013	0.0622	-0.210	83.30 %
Age 25 - 34: sin	-0.026	0.0148	-1.790	7.30 %
Age 25 - 34: cos	0.004	0.0147	0.260	79.20 %
Age 35 - 44: t	0.041	0.0050	8.370	< 0.01 %
Age 35 - 44: growth	0.353	0.0425	8.310	< 0.01 %
Age 35 - 44: unemployment	-0.072	0.0670	-1.070	28.50 %
Age 35 - 44: sin	-0.037	0.0170	-2.170	3.00 %
Age 35 - 44: cos	-0.004	0.0169	-0.210	83.00 %
Age 45 - 54: t	0.019	0.0071	2.730	0.60 %
Age 45 - 54: growth	0.346	0.0603	5.730	< 0.01 %
Age 45 - 54: unemployment	-0.124	0.0927	-1.340	18.10 %
Age 45 - 54: sin	-0.065	0.0244	-2.650	0.80 %
Age 45 - 54: cos	-0.003	0.0242	-0.110	91.30 %
Age 55 - 64: t	-0.064	0.0130	-4.930	< 0.01 %
Age 55 - 64: growth	0.175	0.1054	1.660	9.70 %
Age 55 - 64: unemployment	-0.250	0.1653	-1.510	13.00 %
Age 55 - 64: sin	-0.040	0.0435	-0.920	35.50 %
Age 55 - 64: cos	-0.002	0.0432	-0.040	97.00 %
Age > 64: t	-0.220	0.0477	-4.620	< 0.01 %
Age > 64: growth	0.130	0.3227	0.400	68.70 %
Age > 64: unemployment	0.307	0.4569	0.670	50.20 %
Age > 64: sin	-0.171	0.1303	-1.310	19.00 %
Age > 64: cos	0.030	0.1278	0.240	81.30 %

The *rate* of inflow (the t coefficients in Table 29) of foreign-born Hispanic men to construction is decreasing with age. However, when the age distribution stabilizes (and assuming that the current estimated rates do not change) then the *number* of foreign-born Hispanic men entering construction will be more or less constant up to the age of 55. However, the assumption of stable rates is not likely to hold.

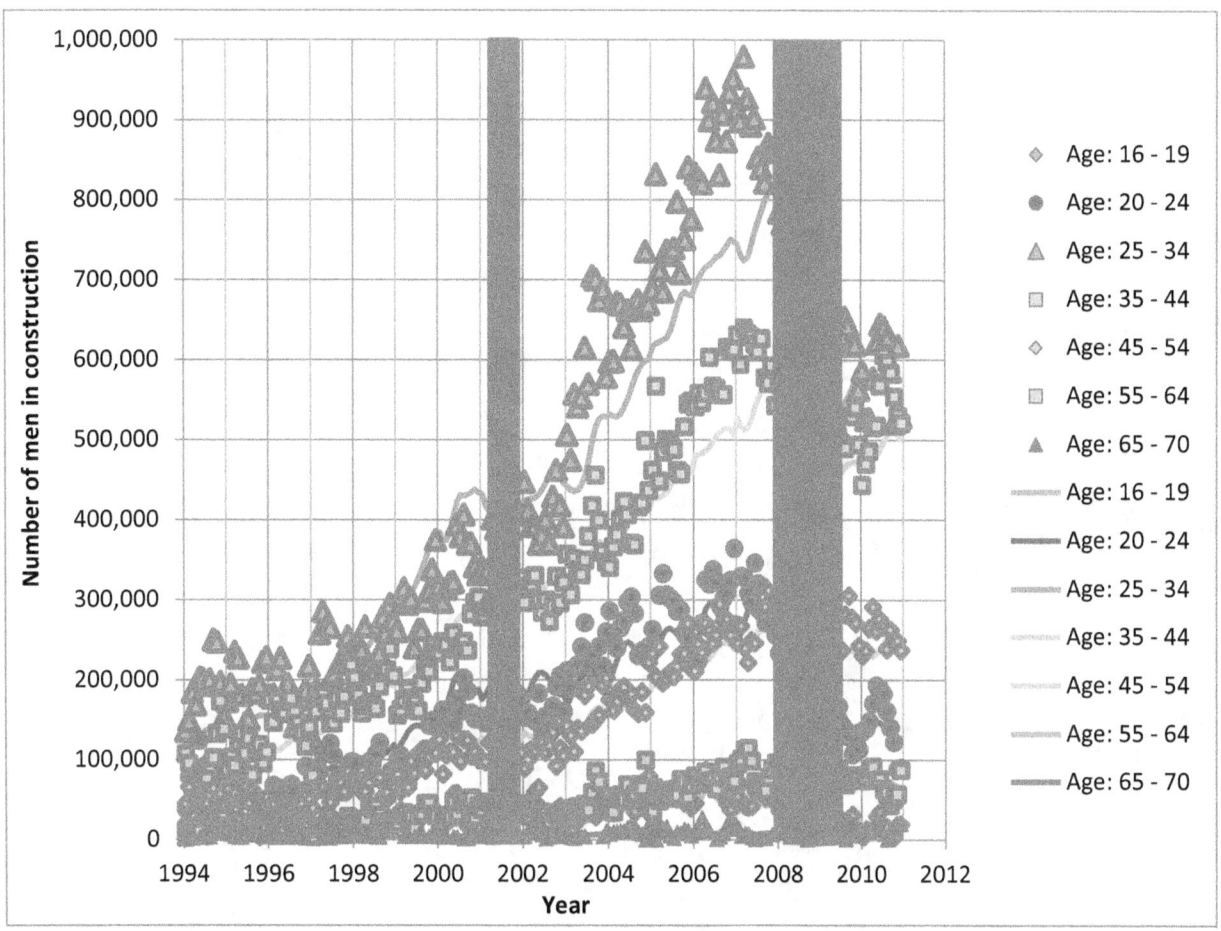

Figure 31: Fit of estimates to data for foreign-born Hispanic men in construction by age group.

It is clear that number of foreign-born Hispanic men in construction is sharply increasing. Net inflow / outflow by age is shown in Figure 32 for selected years. In the average year, 75 % of the foreign-born Hispanic men who enter construction do so before the age of 34. Total net inflow across all age groups is positive in all but three years (data not shown).

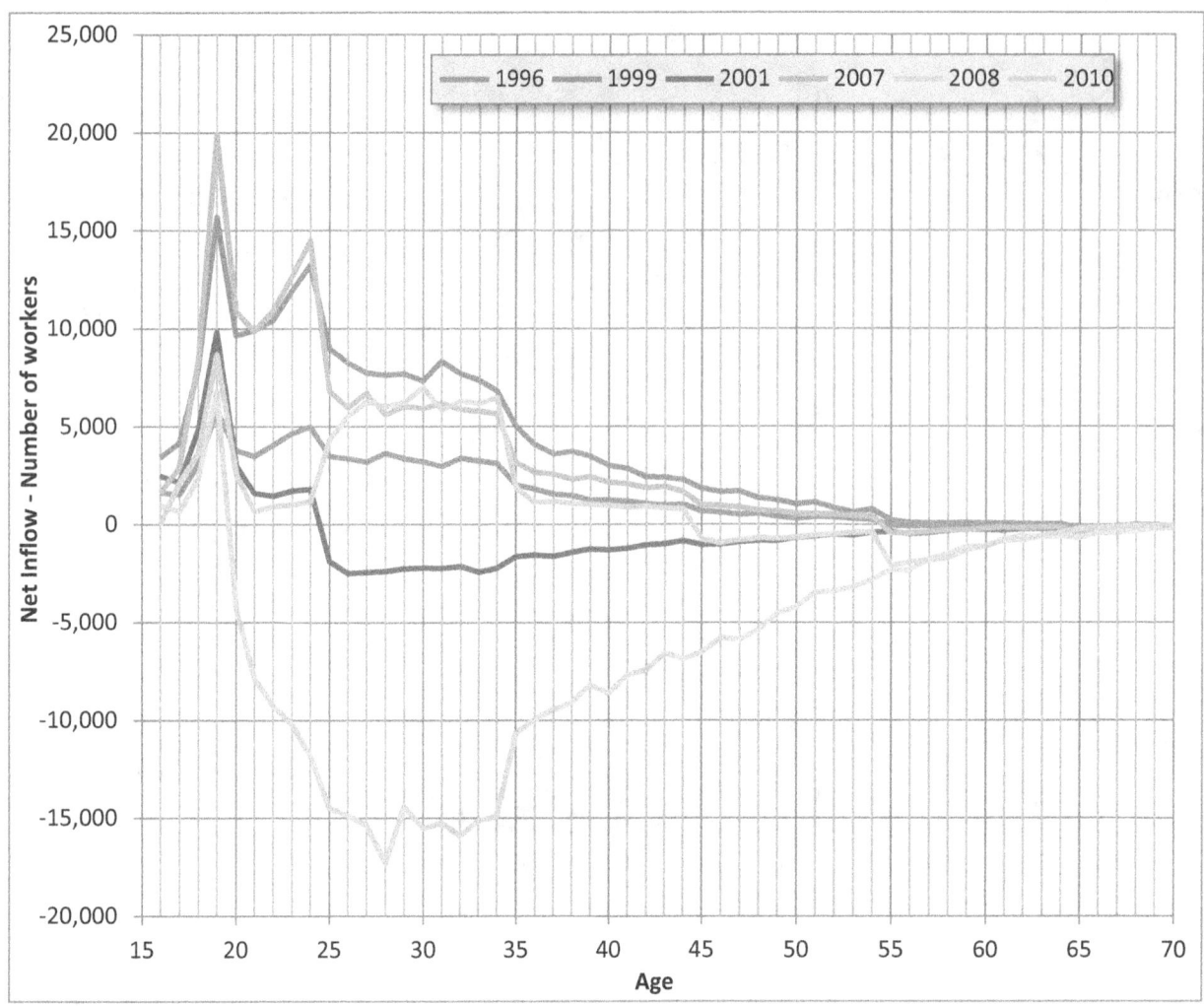

Figure 32: Net inflow to construction for foreign-born Hispanic men, by age for selected years.

Net rate of inflow by age and rate of economic growth is shown in Table 30. As before, this underestimates the variation due to the business cycle, since it does not take into account the general unemployment rate. As usual, older workers are less susceptible to the business cycle than younger workers. Above the age of 55, the impact of the business cycle is not significantly different from zero. Note that it is not until rates of economic growth are much less than average that net inflows turn negative for most age groups.

Impact of seasonality is shown in Table 31. As expected, the impact of seasonality decreases as workers get older, reaching a minimum at around age 35. However, seasonality is not significantly different from zero for any age group except those under age 25. The date when number of workers is highest is concentrated in fall. A Wald test on date of maximum employment indicates that the dates are not significantly different across age groups ($\chi^2 = 1.18$, df = 6, p = 97.77 %).

Table 30: Net rate of inflow to construction by age and rate of economic growth. Colored cells indicate ages and circumstances where men are on net leaving construction.

Age	Deviation from Mean Rate of economic growth						
	-4 %	-3 %	-2 %	-1 %	0 %	1 %	2 %
16 - 19	49.02 %	64.76 %	81.98 %	100.80 %	121.34 %	143.74 %	168.16 %
20 - 24	-13.33 %	-6.08 %	1.70 %	10.03 %	18.95 %	28.50 %	38.70 %
25 - 34	-16.14 %	-10.29 %	-4.10 %	2.44 %	9.36 %	16.67 %	24.39 %
35 - 44	-12.34 %	-8.40 %	-4.32 %	-0.11 %	4.24 %	8.73 %	13.37 %
45 - 54	-13.93 %	-10.15 %	-6.25 %	-2.22 %	1.94 %	6.24 %	10.67 %
55 - 64	-13.93 %	-12.04 %	-10.12 %	-8.18 %	-6.22 %	-4.24 %	-2.24 %
65 - 70	-24.72 %	-23.49 %	-22.26 %	-21.01 %	-19.76 %	-18.51 %	-17.24 %

Table 31: Impact of seasonality. Seasonal Coefficient is the absolute magnitude of the seasonal variation. Wald Test represents the probability that the seasonal variation is greater than zero.

Age	Seasonal Coefficient	% change: lowest to highest	Wald test			Max Date
			χ^2	df	p	
16 – 19	0.2925	79.51 %	8.140	1	0.43 %	13-Apr
20 – 24	0.0883	19.31 %	3.620	1	5.71 %	22-May
25 – 35	0.0268	5.50 %	0.820	1	36.49 %	8-Oct
35 – 44	0.0371	7.70 %	1.190	1	27.63 %	24-Sep
45 – 54	0.0647	13.82 %	1.760	1	18.49 %	27-Sep
55 – 64	0.0403	8.39 %	0.210	1	64.34 %	27-Sep
65 – 70	0.1735	41.50 %	0.440	1	50.69 %	9-Oct

5. Discussion

As discussed earlier, there is a perception that the construction industry has difficulty attracting and retaining skilled workers, and as a result faces a shortage of skilled workers. It appears that the workforce is aging, and that few young people are entering the industry. Training programs for skilled craft workers were traditionally funded and administered through unions, and open-shop training programs have tended to be rare[14]. But since unions have been in a long-term decline, it is not clear where new skilled craft workers will come from. The resulting difficulty staffing projects results in increased costs and schedule delays. This problem is exacerbated by a 30-year decline in real construction wages relative to workers in other industries.

This raises a number of economic questions that this report was intended, in part, to address. Basic factual questions about entry of young people into the industry and trends in union membership are addressed in the earlier sections. Questions about skills and wages require more careful analysis.

It is not clear what is meant by a shortage of skilled labor. The construction market and the construction labor market appear to be highly competitive. Unlike many other industries, the construction market is dominated by a large number of small firms. No one firm has a significant fraction of the market. But in competitive markets, shortages are resolved by increases in price. An increasing price reduces demand and attracts additional supply into the market, thus balancing the market. Huang et al. (2009) report that the "[d]ifficulty in staffing projects has resulted in increasing costs and schedule delays." That raises the question of why wages haven't adjusted for the decline. If construction costs (including delay costs) have increased, why haven't wages?

To begin answering this question, two questions are addressed below. First, to what extent can a decline in skills be discerned in the data? Second, how does labor supply adjust to changes in wage?

5.1. Skills

Since the CPS data does not distinguish skilled craft workers from general construction laborers, and does not directly measure skill level, indirect means must be used to estimate the change in skill level of construction workers over time. One approach would be to assess trends in education levels among construction workers.

To determine whether the educational level of construction workers is changing over time, an index of educational level was constructed and graphed versus age for four different years:

---

[14] See Huang, Allison, Robert Chapman and David Butry. *"Metrics and Tools for Measuring Construction Productivity: Technical and Empirical Considerations."* NIST Special Publication 1101.

1996, 2000, 2006 and 2010. In order to ensure that people were not included multiple times in the sample, the sample was restricted to people in the outgoing rotation.

First, the average years completed in school versus age was graphed (Figure 33). The chart seems to indicate that among men aged 25 – 45, average educational level is decreasing over time. Figure 34 shows that average normalized years of education for men in construction at age 30 by year. Again, there seems to be a decrease in educational level over time. However, this chart suggests that the educational level may be inversely related to the business cycle.

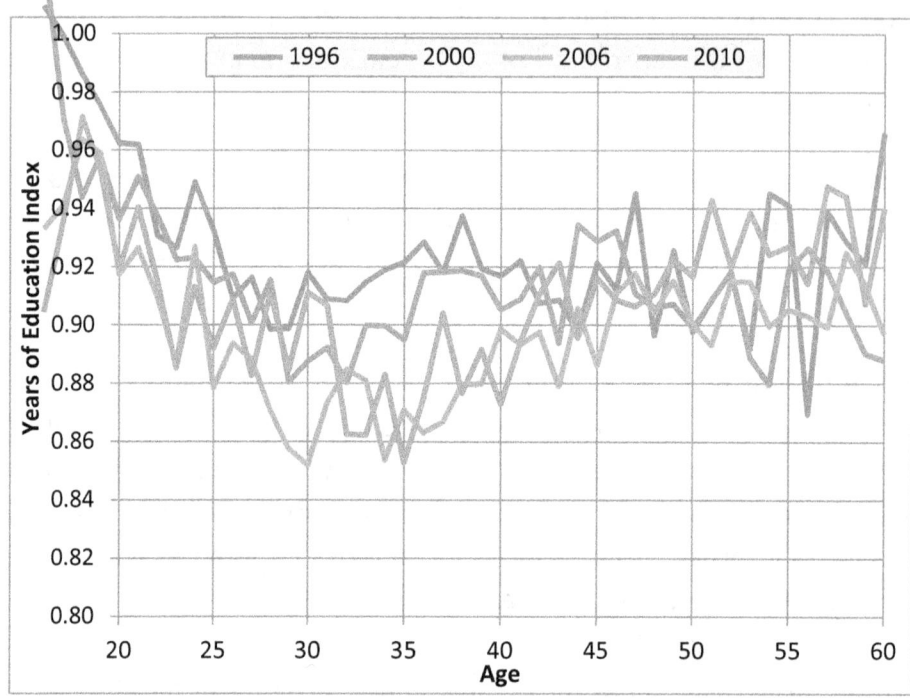

Figure 33: Average years of education of men in construction by age, and normalized by the average years of education of men in the population at large.

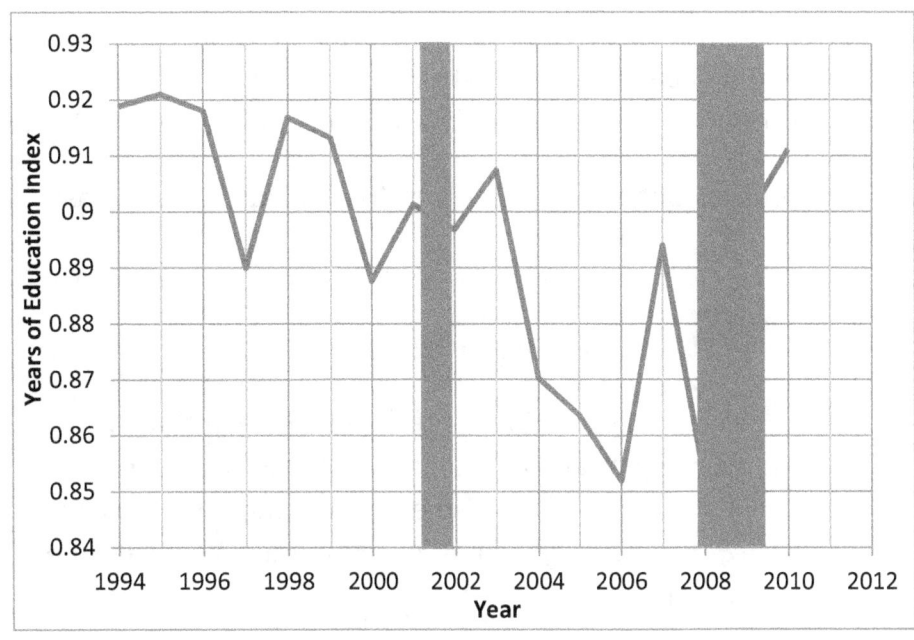

Figure 34: Average normalized years of education for 30-year-old men in construction by year.

Since a formal model of education has not been specified, statistical significance of the observations above cannot be tested. However, this preliminary result provides some support for the idea that there is a decline in skill level among the construction labor force.

5.2. Labor Supply and Wage

The issue of shortages of skilled labor immediately raises questions regarding the nature of supply and demand for construction labor. Preliminary efforts to model supply and demand failed. Specifically, coefficients on price for demand and supply were routinely of the wrong sign. The most likely reason for the failure is that an omitted variable correlates with both wage and employment.

The best candidate for such an omitted variable would be skill. If wages correlate with skill (as seems reasonable) and if low-skill people are the last hired and first fired, then the efforts to estimate supply and demand without taking into account skill will fail. It seems reasonable to assume that skill will correlate with age, education, and years in service. There is no convenient proxy for years in service[15]. Age and education, though, do correlate with wage. However, even after accounting for them the problems with the demand and supply estimations remain.

One way of dealing with this would be to estimate labor supply directly from microeconomic data. As a preliminary approach to doing this, wage and the decision to work in construction

---

[15] One possible proxy would be a simple binary variable identifying whether a person had the same employer upon leaving the survey as when they entered it 16 months earlier (thanks to Dave Butry who suggested it).

for adult men in the CPS were simultaneously estimated. Since the decision to work in construction will be correlated with wage, there will be a selection effect that needs to be accounted for. The basic approach is set forth in Greene, 2000 and Heckman, 1979.

Assume construction wages for any individual are specified as:

$$w_{it} = x_{it} \cdot \beta_w + \xi_t + \varepsilon_{it}$$

where

w_{it} = Construction wage for worker i at time t.
x_{it} = Set of factors that influence the wage received
β_w = Regression parameters associated with wage
ξ_t = fixed-effect variables for average wage at time t
ε_{it} = random variable

The construction wage is only known for people working in construction. If the choice to work in construction were uncorrelated with construction wage, then a simple regression of wage versus regressors would suffice. However, construction wage is a factor in the decision whether to work in construction, so the censoring implicit in the decision to work in construction will bias the regression results. To account for this, a model of employment choice is also specified.

A latent-variable approach is used where:

$$y_{it} = u_{it} \cdot \beta_c + \eta_{it}$$

and worker i works in construction at time t if and only if $y_{it} \geq 0$. Variables are:

y_{it} = Latent variable reflecting whether worker i works in construction at time t.
u_{it} = Set of factors that influence the choice to work in construction
β_c = Regression parameters associated with the choice to work in construction
η_{it} = random variable

Finally, the two error terms are assumed to be jointly normally distributed as:

$$\begin{pmatrix} \varepsilon \\ \eta \end{pmatrix} \sim N\left(0, \begin{pmatrix} \sigma^2 & \sigma\rho \\ \sigma\rho & 1 \end{pmatrix}\right)$$

Note that since the only information we have on y is whether it is greater than or lesser than zero, the initial regression is indeterminiate. By specifying the variance of η as 1, we remove the indeterminacy.

The joint regression is solved by Maximum Likelihood Estimation.

Variables assumed to influence wage include age, age², race/ethnicity, and education. The choice to work in construction is assumed to be influenced by the construction wage, so all the regressors included in the wage equation are included in the choice equation. In addition, choice to work in construction is assumed to be influenced by marital status and whether there are children living at home.

Race/ethnicity, education, marital status, and children are all specified as dummy variables. For race/ethnicity, dummies are included for Black, Native-born Hispanic (Native), and Foreign-born Hispanic (Foreign). White is the omitted category. For education, dummy variables are included for people with less than a high-school education (NoHS) and for people with a college degree (College). The omitted category is people with only a high-school diploma. Marital status (married) is coded for people who are currently married. The Child dummy (childs) is coded if there are children living at home.

Results are shown in Table 32. In the original regression, a fixed-effects variable for each quarter was included, so the parameters represent deviations from the average wage for a given quarter.

Table 32: Simultaneous estimation of construction wage and choice to work in construction for adult men.

	Coef.	Std. Err.	z	P>|z|
wage				
age	38.12563	0.627959	60.71	< 0.001
age2	-0.37647	0.007785	-48.36	< 0.001
Black	-179.542	5.283046	-33.98	< 0.001
Foreign	-161.467	3.752785	-43.03	< 0.001
Native	-72.5199	5.235203	-13.85	< 0.001
College	205.8818	3.097947	66.46	< 0.001
NoHS	-140.119	3.091723	-45.32	< 0.001
construction				
age	0.088357	0.002131	41.47	< 0.001
age2	-0.00112	2.55E-05	-43.87	< 0.001
married	0.210331	0.011291	18.63	< 0.001
childs	0.039352	0.014221	2.77	0.006
Black	-0.25957	0.019085	-13.6	< 0.001
Foreign	-0.03516	0.015884	-2.21	0.027
Native	-0.10151	0.021117	-4.81	< 0.001
College	-0.04251	0.013081	-3.25	0.001
NoHS	-0.10035	0.012251	-8.19	< 0.001
rho	-0.14258	0.013198		
sigma	386.0331	0.824089		

Most variables are pretty much what one would expect with one important exception. In the estimation *rho* is the correlation between the wage and the choice to work in construction otherwise unaccounted for in the model. In this model *rho* is negative. That implies that increases in construction wages are associated with people selecting **out** of construction.

As a general rule one would expect people to gravitate to the higher wage occupation. So if higher construction wages are associated with people selecting out of construction, then higher (potential) construction wages must generally be associated with higher wages in other industries as well. If it is assumed that workers are paid their marginal productivity (i.e., higher wage workers are more productive), then the results imply (all else equal) that workers who are more skilled at construction are also more skilled at other jobs as well. That is unsurprising, but the *negative* correlation between wage and choice of construction implies something stronger: the skill premium is higher in other industries than in construction. So (again, all else equal) there are relatively few 'highly skilled' construction workers because such people can earn more in other industries—presumably because they are more productive there.

This suggests that part of the decline in numbers of skilled craft workers may be attributable to macroeconomic factors. It is well known that the premium to college education has increased since the 1970's (see for example Gottschalk, 1997). This increase in the college premium is partially attributable to a decline in wages for people with only a high-school education. This suggests that the decline in skilled craft workers may be driven by a larger decline in demand for low-skilled labor generally. Since construction seems to gain less in productivity terms from skill and ability than other industries, the growing skills premium in the modern economy induces people with relatively high skill and ability to seek employment elsewhere.

5.3. Future Directions

There are a number of additional directions that would contribute to understanding the construction labor market.

- Characterization at the regional / local level.

    This report characterized labor supply at the national scale for the most part. However, construction is primarily a local market and there will be aspects of the market that will be obscured by looking at it nationally. For example, racial makeup (and probably seasonality) clearly differs from region to region. So deepening the analysis to look at the data at a regional scale would likely improve our understanding of the market.

- Estimation of supply and demand.

    Estimating supply and demand functions for construction labor would help. That turns out to be surprisingly difficult due to the high correlation between wages and employment. During times of increasing employment, wages (presumably) increase, but the people hired are at the low end of the wage scale while the people at the high end of the wage scale are susceptible to poaching by other industries. Times of decreasing employment present the reverse situation. That makes it difficult to tease out the relationship between supply and wage holding all else constant. Completing the task of estimating supply and demand will help fill in some of the missing pieces of the picture of the construction labor market.

- Labor Unions

    Some of the concern about productivity in construction centers on a perceived shortage of skilled labor. That seems to be linked to the decline in union membership. So to understand what is going on with this, an understanding is needed of the place of unions in the market, why market share is declining (both from the supply side with people choosing whether to join and from the demand side of builders choosing whether to hire union labor), and what (if anything) is replacing unions in the marketplace. So to better understand the nature of skilled-labor shortages (or lack thereof) requires an understanding of the changing place of the trade unions in the market.

- Wage trends for skilled craft workers versus general construction labor

    The data above, derived from the CPS, is not detailed enough to distinguish skilled craft workers in construction from general laborers. It is possible that average wages are declining while wages for skilled craft workers are increasing. This would explain the conundrum of "shortages" of skilled craft workers alongside declining construction wages. One way of assessing this possibility would be to look at the long-term trend of wages for craft workers versus general construction labor. Such data does not exist in the CPS, so other data sources would have to be found for such information.

- Labor flows by educational level

    Expanding the analysis of labor flows to address educational levels would provide additional insight into long-term changes in educational levels in people entering construction.

- Analysis by market segment

    Eventually, this work needs to be done for different segments of the construction market. Housing is such a large portion of the market that the results above are likely dominated by that segment of the market. But other segments will likely be different.

6. References

Gottschalk, P. (1997, Spring). Inequality, Income Growth, and Mobility: The Basic Facts. *The Journal of Economic Perspectives, 11*(2), 21-40.

Greene, W. (2000). *Econometric Analysis* (4th ed.). Upper Saddle River, NJ: Prentice Hall.

Heckman, J. (1979). Sample Selection Bias as a Specification Error. *Econometrica, 47*(1), 153-161.

Huang, A., Chapman, R., & Butry, D. (2009). *Metrics and Tools for Measuring Construction Productivity: Technical and Empirical Considerations*. Gaithersburg, MD: National Institute of Standards and Technology.

StataCorp. (2009). *Stata Statistical Software: Release 11*. College Station, TX.

Teicholz, P. (2004, 14 April). Labor Productivity Declines in the Construction Industry: Causes and Remedies. *AECbytes Viewpoint*(4).

Thomas, D. (2010). *Methodology for Calculating Construction Industry Supply Chain Statistics*. Gaithersburg, MD.: National Institute of Standards and Technology.

Appendices

Appendix 1: Data

Below is a table of variables used from the Current Population Survey

Variable	Period	Description
HRHHID	1994-2010	Household ID
HRHHID2	2004-2010	Part 2 of the Household ID Number (= the next 3 variables concatenated)
HRSAMPLE	1994-2004	Sample Number
HRSERSUF	1994-2004	Serial suffix
HUHHNUM	1994-2004	Household Number (replacement households)
OCCURNUM	1994-2010	Person Number in Household (= PULINENO)
HRMIS	1994-2010	Month in Sample (1 – 8)
HRMONTH	1994-2010	Month
HRYEAR	1994-1998	2-Digit Year
HRYEAR4	1998-2010	4-Digit Year
GEREG	1994-2010	Region (1 – Northeast, 2 – Midwest, 3 – South, 4 – West)
PESEX	1994-2010	1 – Male
PRTAGE	1994-2010	Age as of the end of the survey week
PERACE	1994-2002	Race (see table in Codebook) (This is recoded in 1997)
PTDTRACE	2003-2010	Detailed Race (see table in Codebook)
PRHSPNON	1994-2002	Hispanic Status
PEHSPNON	2003-2010	Hispanic Status
PRCITSHP	1994-2010	Citizenship Status (1-3 – Native, 4 – Naturalized, 5 – Non Citizen)
PEEDUCA	1994-2010	Educational Levels (see Codebook for definitions)
PEMLR	1994-2010	Labor Force Status (see table in Codebook)
PEHRUSL1	1994-2010	Usual hours worked at job 1 (-4: hours vary)
PEHRACT1	1994-2010	Actual hours worked at job 1 last week.
PEERNHRY	1994-2010	Indicator for Hourly Employee. Only sampled when HRMIS $\in \{4, 8\}$
PEERNLAB	1994-2010	Indicator for Labor Union Member. Only sampled when HRMIS $\in \{4, 8\}$
PEIO1COW	1994-2010	Class of Worker (1st job) (Gov't; Private; Self-Employed; unpaid)
PRDTIND1	1994-2010	Detailed Industry, Job 1. Definitions changed at the end of 2002.
PRDTOCC1	1994-2010	Detailed Occupation, Job 1. Definitions changed at the end of 2002.
PUIODP1	1994-2010	Same Employer (1: Yes, 2: No, < 0: No Answer)
PTERNWA	1994-2010	Weekly Earnings (to the penny)
PWORWGT	1998-2010	Outgoing Rotation Weight
PWLGWGT	1994-2010	Longitudinal Weight
PWSSWGT	1994-2010	"Six-Step" Weight

The CPS independently asks people what industry they work in and whether they have changed jobs in the previous month. That sets up a quality check of the data. Specifically, there should be very few people who have changed industries without changing jobs. However, before the middle of 1995 approximately 15,500 people per month are listed as changing industries without changing jobs. That suggests that the data before the middle of 1995 may not be as reliable as the rest of the data.

Time Period	Average	Max	Total
Jan '94 – Jun '95	15,561	30,501	278,293
Jul '95 - 2010[16]	252	632	45,399

---

[16] Excluding the end of 2002 when industry definitions changed.

Other results give a similar impression.

Data for quarterly GDP were obtained from the Bureau of Economic Analysis. Monthly data for Unemployment and Urban Consumer Price Index (CPI-U) were obtained from the Bureau of Labor Statistics. This reported used data for 1993 – 2010.

Since the main data are monthly and GDP are published quarterly, a method was needed to interpolate the GDP data to monthly to effectively make use of it. To interpolate the data two assumptions were made:

1. Reported quarterly growth is the geometric mean of the interpolated monthly growth rates for the months making up the calendar quarter. That is (for example), the geometric mean of growth for April, May and June 2007 equals the reported growth for the second quarter 2007.

2. Log of monthly growth is linearly interpolated between the log growth rates of the central months in each quarter. That is (for example), March 2007 and April 2007 growth are linearly interpolated between Feb 2007 and May 2007.

If t represents months, g_t represents the published growth rate for a calendar quarter (where the published growth rate is arbitrarily assigned to the central month in the quarter), and $\tilde{g}_t$ represents interpolated monthly growth rate, then assumption 1 above is specified as:

$$\frac{1}{3}\ln(1+\tilde{g}_{t-1}) + \frac{1}{3}\ln(1+\tilde{g}_t) + \frac{1}{3}\ln(1+\tilde{g}_{t+1}) = \ln(1+g_t)$$

For the last month in a calendar quarter, assumption 2 is specified as:

$$\frac{2}{3}\ln(1+\tilde{g}_{t-1}) - \ln(1+\tilde{g}_t) + \frac{1}{3}\ln(1+\tilde{g}_{t+2}) = 0$$

And similarly for the first month in a calendar quarter.

That produces a set of simultaneous equations, with two more unknowns than equations. By specifying that Jan 1993 and Dec 2010 are linear extensions of the subsequent (previous) series of months, there is a fully specified set of linear equations. Note that GDP is reported as annualized growth rates, and the equations listed above preserve that convention for the interpolated monthly growth rates. A graph of the interpolated monthly GDP figures overlaid on the reported quarterly GDP figures is shown in Figure 35.

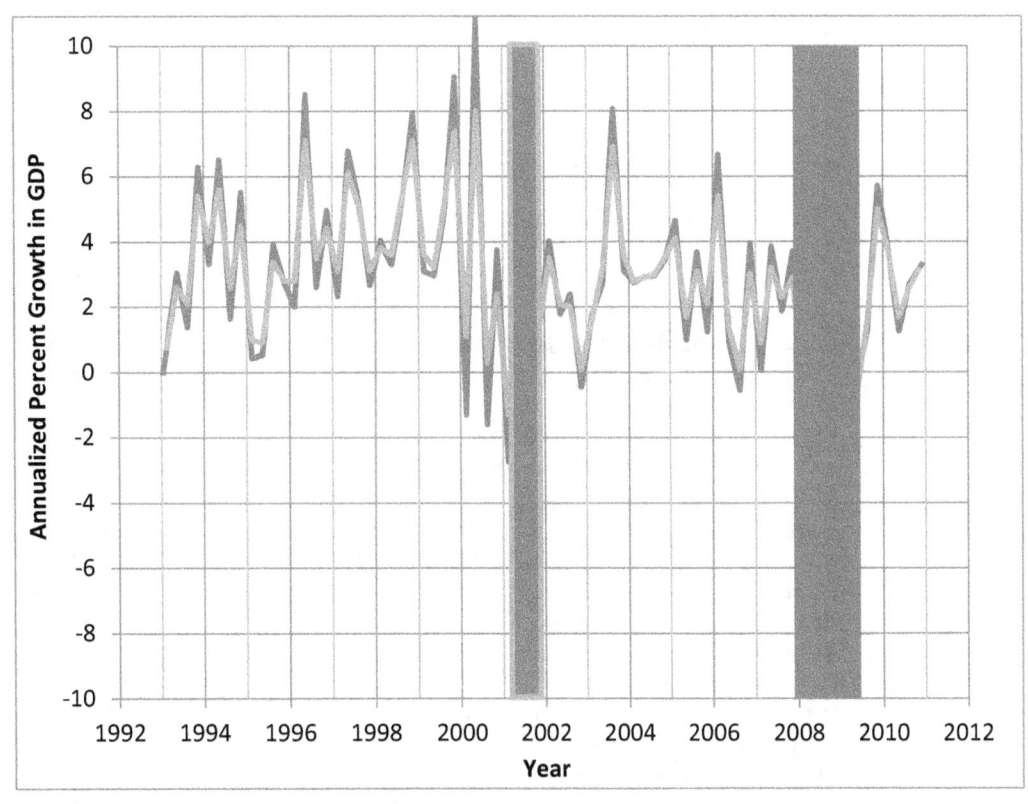

Figure 35: Interpolated Monthly GDP overlaid on reported Quarterly GDP.

Appendix 2: Similarity Indices

Industry	r_{ij}	f
Agriculture	12.60 %	2.25 %
Forestry, logging, fishing, hunting, and trapping	16.16 %	0.34 %
Mining	15.12 %	0.91 %
Construction	100.00 %	100.00 %
Nonmetallic mineral product manufacturing	14.24 %	0.74 %
Primary metals and fabricated metal products	10.11 %	2.23 %
Machinery manufacturing	7.08 %	0.92 %
Computer and electronic product manufacturing	3.24 %	0.38 %
Electrical equipment, appliance manufacturing	3.18 %	0.21 %
Transportation equipment manufacturing	4.30 %	1.43 %
Wood products	17.04 %	0.71 %
Furniture and fixtures manufacturing	14.70 %	0.86 %
Miscellaneous and not specified manufacturing	4.73 %	0.59 %
Food manufacturing	5.84 %	0.97 %
Beverage and tobacco products	4.44 %	0.18 %
Textile, apparel, and leather manufacturing	3.22 %	0.30 %
Paper and printing	3.83 %	0.49 %
Petroleum and coal products manufacturing	15.45 %	0.32 %
Chemical manufacturing	5.25 %	0.64 %
Plastics and rubber products	3.89 %	0.55 %
Wholesale trade	6.65 %	3.09 %
Retail trade	5.02 %	8.93 %
Transportation and warehousing	6.60 %	4.66 %
Utilities	10.57 %	1.60 %
Publishing industries (except internet)	0.97 %	0.17 %
Motion picture and sound recording industries	6.18 %	0.19 %
Broadcasting (except internet)	9.65 %	0.57 %
Internet publishing and broadcasting	0.00 %	0.00 %
Telecommunications	2.22 %	0.45 %
Internet service providers and data processing services	3.41 %	0.07 %
Other information services	0.00 %	0.01 %
Finance	1.38 %	0.83 %
Insurance	1.87 %	0.51 %
Real estate	7.02 %	1.82 %
Rental and leasing services	6.13 %	0.30 %
Professional and technical services	4.00 %	3.24 %
Management of companies and enterprises	0.34 %	0.00 %
Administrative and support services	13.80 %	6.61 %
Waste management and remediation services	14.85 %	0.65 %
Educational services	1.84 %	2.52 %
Hospitals	1.15 %	0.89 %
Health care services, except hospitals	1.27 %	1.15 %
Social assistance	0.99 %	0.37 %
Arts, entertainment, and recreation	6.02 %	1.50 %
Accommodation	4.72 %	0.66 %
Food services and drinking places	5.87 %	4.52 %
Repair and maintenance	13.61 %	2.81 %
Personal and laundry services	4.03 %	0.76 %
Membership associations and organizations	3.03 %	0.60 %
Private households	5.22 %	0.35 %
Public administration	2.82 %	2.05 %

www.ingramcontent.com/pod-product-compliance
Lightning Source LLC
Chambersburg PA
CBHW081730170526
45167CB00009B/3774